Comparing Taliban Social Media Usage by Language

Who's Speaking and What's Being Said

BRADLEY M. KNOPP, JON NIEWIJK, ZOHAN HASAN TARIQ,
ELMO C. WRIGHT, JR.

Approved for public release; distribution unlimited

RAND NATIONAL DEFENSE RESEARCH INSTITUTE

For more information on this publication, visit **www.rand.org/t/RRA1830-1**.

About RAND

The RAND Corporation is a research organization that develops solutions to public policy challenges to help make communities throughout the world safer and more secure, healthier and more prosperous. RAND is nonprofit, nonpartisan, and committed to the public interest. To learn more about RAND, visit www.rand.org.

Research Integrity

Our mission to help improve policy and decisionmaking through research and analysis is enabled through our core values of quality and objectivity and our unwavering commitment to the highest level of integrity and ethical behavior. To help ensure our research and analysis are rigorous, objective, and nonpartisan, we subject our research publications to a robust and exacting quality-assurance process; avoid both the appearance and reality of financial and other conflicts of interest through staff training, project screening, and a policy of mandatory disclosure; and pursue transparency in our research engagements through our commitment to the open publication of our research findings and recommendations, disclosure of the source of funding of published research, and policies to ensure intellectual independence. For more information, visit www.rand.org/about/principles.

RAND's publications do not necessarily reflect the opinions of its research clients and sponsors.

Published by the RAND Corporation, Santa Monica, Calif.
© 2023 RAND Corporation
RAND® is a registered trademark.

Library of Congress Cataloging-in-Publication Data is available for this publication.
ISBN: 978-1-9774-1078-8

Cover image: Muhammet Camdereli/Getty Images

About This Research Report

In the wake of the Taliban takeover and the reestablishment of the Islamic Emirate of Afghanistan in August 2021, there are significant questions about the best methods to interpret Taliban messaging. Better interpretation would, in turn, help U.S. and Western policymakers to better understand Taliban leadership intentions and would thus allow for better-informed U.S. and Western policy decisions regarding Afghanistan. Limited direct diplomatic contact between the Taliban and most Western governments necessitates reliance on alternative sources of information. Taliban use of social media, particularly Twitter, was identified as one way to understand Taliban thinking. Previous studies of Taliban use of social media noted historical differences between Taliban messaging to the West and Taliban messaging to regional audiences, and those studies identified this difference as a potential way to inform Western policy actions. In this report, we examine Taliban leadership use of Twitter in various regional languages, and we seek to clarify the posts being broadcast through this medium and to determine whether messaging in the various languages is similar or different.

National Security Research Division

This research was conducted within the International Security and Defense Policy Program of the RAND National Security Research Division (NSRD), which operates the RAND National Defense Research Institute (NDRI), a federally funded research and development center (FFRDC) sponsored by the Office of the Secretary of Defense, the Joint Staff, the Unified Combatant Commands, the Navy, the Marine Corps, the defense agencies, and the defense intelligence enterprise. This research was made possible by NDRI exploratory research funding that was provided through the FFRDC contract and approved by NDRI's primary sponsor.

For more information on the RAND International Security and Defense Policy Program, see www.rand.org/nsrd/isdp or contact the director (contact information is provided on the webpage).

Acknowledgments

We thank Mike McNerney for urging us to take on this fascinating research task and for providing his support. We also acknowledge the contributions made to this report by Elizabeth Hammes, who provided critical early research services for the team. Michael Ryan provided irreplaceable support concerning collection of the Twitter data used in this work; we thank him also for his patience in providing assistance with Brandwatch, which helped to arrange and analyze the Twitter data. For their critical reviews of the report, we thank Jeff Martini and Laurel Miller of RAND; their contributions improved the work.

Summary

We studied Taliban use of social media messaging to gain insight regarding key personnel, policies, and themes and to examine whether the media messaging across various regional languages was similar or different. We assessed that understanding any differences in messaging between foreign and domestic audiences would increase understanding of Taliban policies and policymaking. Our study focused on Twitter, which is the most broadly used social media platform to publicize news and policy by Taliban leaders and spokesmen and provided us with the largest and most manipulable database to support our research. Our team amassed a large collection of tweets for analysis spanning the period from the Taliban's takeover of Afghanistan in August 2021 to April 2022. These tweets are captured in a database for further analysis and review.

When we looked at specific thematic areas, we found many similarities but some distinct differences in Taliban messaging among the various languages and target audiences and on different topics. We focused on three areas: economic concerns, relationships with other militant groups and neighboring countries, and the status of women in Afghan society.

We documented that the Taliban leadership composition has remained in the same hands, in some cases passing from father to son, since the group's inception in the 1990s. Those in the top echelon are exclusively male, Pashtun, graduates of Pakistan seminaries, sanctioned by the United Nations, and (to some extent) former Guantanamo detainees.

Concerning the economy, the Taliban appear to be attempting to craft a narrative to dodge responsibility for the 2022 economic crisis by building the case that, because of rampant corruption, the economy was crumbling under the previous regime.

Overall, Taliban members use strong negative language when addressing the public about the U.S. presence in Afghanistan. In general, messaging about the United States is negative, although some coverage of meetings with the U.S. Special Representative has sounded a positive note. The Taliban media team extensively covers foreign visits by Taliban officials and meetings with foreign dignitaries at home and abroad, but the messaging is often procedural. On occasion, some spokespersons offer comments on events in the world that affect Afghanistan, but this is mostly in the shape of sharing the Islamic Emirate of Afghanistan (IEA)'s official statement on the matter. Interactions with China, Russia, Turkmenistan, and Uzbekistan are robustly covered, but references to the third northern neighbor, Tajikistan, are almost nonexistent. Relations with Pakistan, which has not recognized the Taliban government, are cordial but some tensions exist. There are virtually no contacts with India.

The Taliban are known to have relationships with a few militant groups, such as al Qaeda, the Tehrik-e-Taliban Pakistan (TTP), the Islamic Movement of Uzbekistan, and the Turkistan Islamic Party. Since August 2021, no Taliban spokesperson has mentioned any of these groups by name, though the Taliban have overtly supported talks between the TTP and Pakistan. Islamic State Khorasan Province (ISKP), the Islamic State's regional branch, has posed the toughest security challenge to the Taliban since the establishment of the Islamic Emirate in August 2021. Taliban messaging about ISKP is limited but mixed.

Messaging about women's rights has been prominent in Taliban tweets in Pashto, Dari/Farsi,[1] Arabic, and English. However, the mix of themes surrounding women's issues that are discussed varies across language groups. The most-notable differences were Arabic messaging's relatively low emphasis on the IEA's support for women's education and workforce participation compared with English messaging's relatively low emphasis on aligning women's rights with Islamic values.

Our study did not discern an overall strategy behind Taliban social media use; there is no strong evidence that Taliban social media activity is coordinated at a high level. Individual leaders use social media to inform domestic and regional constituencies about issues for which the leaders are responsible. Our analysis suggests that the posts are designed to inform, persuade, and influence a target audience. Taliban messaging on some issues was common across languages; in other cases, the messaging was distinctly different, reflecting Taliban interest in tailoring a post for a specific audience without regard to potential contradictions by other individuals or among the various languages. Our analysis reinforced the hypothesis that understanding the differences in Taliban messaging for internal and external audiences can provide valuable insights into what Taliban leaders are saying on key issues, what messaging they want to convey, and how they wish to shape those policy issues. An area of particular interest in our analysis is comparing policy implementation with the frequency of posts on a given topic (as an indicator of the issue's importance to the Taliban) and the aspiration articulated in that messaging and the actual implementation of the policy in question.

The identification of important differences in Taliban messaging shows the value of monitoring that messaging in different regional languages for both content and tone. Use of a similar analytic approach on other important issues might provide similarly useful insights. The recommendations listed here represent follow-on questions that would continue to inform U.S. and Western policy deliberation regarding the IEA:

- Consider expanding the analysis of differences between English messaging and messaging in regional languages to encompass other countries (such as Pakistan, India, and Iran) and other militant groups (such as ISKP) to broaden understanding of the wider regional situation.
- Consider applying a similar approach in monitoring both (1) the anti-Taliban (former Northern Alliance) groups and their military activities in the north and (2) former national security forces that were trained by the North Atlantic Treaty Organization.

---

[1] There is a dispute in Afghanistan over the name of the official language. Along with Pashto, Dari is one of the two official languages of Afghanistan. However, many Dari speakers refer to their language as *Farsi* and argue that the name *Dari* was forced on them by the dominant Pashtun ethnic group. To avoid the appearance of taking sides on this issue, we use the term *Dari/Farsi* throughout the paper.

- Consider establishing a full-time team of native linguists and analysts with area expertise to continue to monitor Taliban messaging and provide in-depth, culturally sensitive insight into longer-term Taliban goals and actions and into leadership dynamics within the group. Close monitoring by native linguists of these social media activities could provide policymakers with a nuanced cultural appreciation for events; this, in turn, could provide insights regarding domestic and regional developments.

Contents

Tables

CHAPTER 1

Introduction

Taliban use of social media has evolved over the decades of the movement. The use of social media was forbidden when the Taliban governed Afghanistan from 1996 until late 2001. After they were ousted from power in December 2001, Taliban use of social media grew significantly. Their embrace of social media during the 2000s was driven by several needs: to recruit new members, to amplify their messaging to the international community, and to communicate their positions and aspirations throughout Afghanistan. Since taking over the government in August 2021 after the departure of the U.S.-led military forces, the Taliban have used social media more robustly, notably Facebook and Twitter:

- Facebook is available to individual users in Afghanistan, but the Taliban as an organization remain prohibited from establishing and using the app. Facebook stated that it would continue its ban on the Taliban, which Facebook defines as a terrorist organization.[1]
- Twitter, on the other hand, continues to host accounts by senior Taliban leaders. Twitter claims that it applies its content moderation policies with regard to violent, hateful content and abusive behavior. Both official and personal accounts are used by the Taliban on Twitter. There is some evidence of Twitter accounts set up to propagate messaging, but the extent of this practice is unclear.

The relatively quick embrace and continued use of social media by the Taliban (now that they are back in power) raises several important questions:

- Do the Taliban use or plan to use social media, especially Twitter, in an organized, strategic way to communicate domestically and with regional and international partners?
- Which social media platforms, and which accounts on those platforms, are the most important?
- Are posts in the different regional languages—Pashto, Dari/Farsi, Arabic, and Urdu—communicated similarly? How do those posts compare with the posts that the Taliban is sending in English to Western audiences?

Addressing these questions will provide insight into current and future Taliban actions, especially those that might affect U.S. and Western interests.

---

[1] Meta, "Dangerous Individuals and Organizations," webpage, undated.

Methodology

Our multilingual team identified a set of key leaders and influencers who regularly used social media, specifically Twitter, to communicate policy. Research was conducted on tweets in Pashto, Dari/Farsi, Arabic, Urdu, and English. Our research is based on an analysis of Taliban-affiliated posts on Twitter, the social media platform used most frequently by Taliban leaders and spokesmen to publicize news and policy. Tweets collected from select Twitter accounts provided our research team with the largest and most manipulable database to support our research. Facebook and other major social media players, such as YouTube and TikTok, do not allow official Taliban accounts on their sites. The Taliban established an official website in 2005.[2] The site is in five languages (Pashto, Dari/Farsi, Arabic, Urdu, and English), but it is not updated regularly and the content is inconsistent among the different language sites. Other tools, such as WhatsApp, are used by the Taliban, but these applications have limitations that lessen their effectiveness for broad dissemination internally and externally. Large swaths of the Afghan public do not have computer access and instead use smartphones for internet connectivity, so the website is not as agile as Twitter, which operates via phone and can quickly circulate communication domestically and internationally—a fact that also underpinned our decision to focus on that particular app.[3]

Twitter also operates in all the regional languages and thus provides the broadest snapshot of Taliban communications. Several considerations led us to selecting the four regional languages for analysis. Pashto is the dominant language used in Taliban messaging on Twitter. This reflects the dominance of the Pashtun tribal confederation in the Taliban leadership. Dari/Farsi is the language more broadly spoken across Afghanistan and provides the next largest portion of posts we examined. Urdu was also used but to a lesser degree and mostly addresses the government and people of Pakistan. Arabic is also used to a lesser degree, but we decided to use Arabic tweets in our analysis because this is the language used by Taliban negotiators in Doha. We hypothesized that individuals in Doha would be tweeting in Arabic seeking to influence Arabic-speaking populations across the Middle East. Finally, English is the language that reaches most other international audiences and is the language most frequently used to address foreign audiences, especially in the West. The following list shows the distribution of languages spoken in Afghanistan (the overlap in percentages represents individuals who speak or understand several of the regional languages):

- Dari/Farsi: 77 percent
- Pashto: 48 percent
- Uzbek: 11 percent

---

[2] *Voice of Jihad* (Al Emarah English), homepage, undated. Each language has a separate website.

[3] Only around 23 percent of Afghanistan's population has internet access while around 69 percent of the population have mobile connections. DataReportal, "Digital 2022: Afghanistan," webpage, February 15, 2022.

- English: 6 percent
- Urdu: 3 percent
- Arabic: 1 percent.[4]

Assessing Twitter posts in these languages provided us with a different view into Taliban thinking. Because Al Emarah maintains these five languages on its website, we were able to compare our assessments of Twitter communications with our assessments of material from an official Taliban media outlet. We did not address Uzbek or other languages of populations in Afghanistan, largely because our initial review of Taliban messaging did not reveal significant numbers of posts in these languages but also because we assessed that the most-influential individuals would use one or more of the four languages we did analyze. In addition to analyzing Twitter posts, we reviewed news reports in local languages and in English and official Taliban statements made at public venues (such as international conferences). The purpose was to compare the Taliban's stated policy intent with actual actions they took and, in some cases, to provide additional context to the tweets.

Our database of Twitter posts was collected via the RAND Corporation's account with Twitter. Posts were compiled into Excel spreadsheets and coded according to date, originator, language, and topic.

At the time of our research between August 2021 and April 2022, all these tweets were still available online. Some have since been deleted or archived. The references at the end of this report provide links to the tweets cited in the study that were still available at the time of publication.

How We Selected Twitter Accounts to Analyze

To select the Twitter accounts for analysis, we focused on leaders, known policymakers, and influencers inside and outside Afghanistan to establish a broad cross section of individuals whose positions or reputations for influence could be considered to represent official Taliban policies and thinking. We initially used the list of individuals who were announced as leaders of the interim government shortly after taking control of the country, but several of these individuals did not have Twitter accounts. To expand the size of the group to be surveyed, we added individuals who frequently spoke on behalf of Afghan leaders, including those leaders with no individual Twitter accounts. We also identified several active institutional Twitter accounts, including those of the Ministry of Foreign Affairs (MoFA) and the Ministry of Defense. Other accounts, such as the Taliban representatives in Doha, were identified by frequent use and policy content.

Twitter data for this report were in some cases supplemented with relevant materials from other news outlets, mainly local media, to contextualize and validate information in the tweets. Several Taliban officials and spokespersons actively use Twitter; some of them have

---

[4] Central Intelligence Agency, "Afghanistan," World Factbook, May 31, 2022.

hundreds of thousands of followers. All combined, they tweet in four languages—Pashto, Dari/Farsi, Arabic, and English—with most tweets in the local languages (Pashto and Dari/Farsi). Some of these Taliban members are polyglots and post the same content in multiple languages. The Taliban Twitter posts cover a broad variety of topics pertaining to governance, security, economics, education, engagement with other countries, Cabinet meetings, appointments of senior officials, and domestic and international trips of Taliban members. The major themes are synopses of meetings with, interviews with, and speeches delivered by Taliban leaders and cabinet members. Another major theme is decrees issued by their supreme leader.

Twenty Twitter accounts affiliated with the Taliban were examined for this study. The users include Taliban spokespersons; members of the political office in Doha, Qatar; MoFA; the Office of Deputy Prime Minister; and individuals close to the Taliban (see Table 1.1).[5] Taliban messaging is decentralized. Various departments focus on different topics given their scope of work, but the posts also accentuate their faction's preferred approaches to governance, economy, women's rights, engagement with the international community, and desire to aim messaging at different audiences. For instance, the members of the Haqqani network are silent about girls' education but actively praise Jihad and martyrdom. Conversely, the members of political office in Doha, where the Taliban have maintained an office since 2013 and where the negotiations with the United States occurred, try to paint a moderate picture of the Taliban regime that respects women's rights and supports girls' education. Because of this decentralization, our team selected this broader variety of individuals and used sampling criteria to capture a greater variety of policy messaging. Serving in the Taliban de facto government, being closely affiliated with the Taliban but without an official appointment, and tweets reflecting the policies of the administration were the main sampling criteria.

Throughout the report, the individual tweets referenced in the text are included in their entirety in the associated footnotes.

All the tweets posted by these 20 users were collected using an algorithm that automated the data collection processes. We aimed to analyze the tweets posted from August 15, 2021 (the date of the fall of Kabul to the Taliban and the start date of the Taliban de facto government), through April 15, 2022. The Taliban posts made prior to August 15, 2021, were predominately about war. Since taking power, Taliban messaging has focused on governance, the economy, and security.[6] The final dataset analyzed for our study included a total of 15,157 tweets. Table 1.2 illustrates the language distribution of the tweets in the final dataset.

---

[5] The Taliban leadership members, such as the prime minister and his two deputies, do not have individual accounts on Twitter, but there is one account affiliated to the office of First Deputy Prime Minister Abdul Ghani Barader.

[6] As noted in Table 1.2, the start dates and end dates differ for each account. The start dates are different because some of the accounts were created after August 15, 2021. The end dates vary because we collected and cleaned the data of each account separately.

TABLE 1.1
Twitter Accounts Surveyed

Name	Position	Start Date	End Date	Tweet Count	Total Followers
		Data Collection			
Atiqullah Aziz	Deputy Minister for Culture and Arts	August 17, 2021	April 2, 2022	139	9,929
Qari Yousaf Ahmadi	Former Taliban spokesperson	August 15, 2021	March 21, 2022	87	116,500
Zabeh Ullah	Deputy Chief of Staff of 1st Deputy Prime Minister	September 13, 2021	March 23, 2022	19	3,589
MoFA	Official Twitter account	December 23, 2021	March 31, 2022	383	11,700
Hafiz Zia Ahmad	Deputy Spokesman and Assistant Director of Public Relations, MoFA	December 7, 2021	April 5, 2022	264	3,983
Office of Mullah Ab Ghani Barader	First Deputy Prime Minister	November 14, 2021	March 31, 2022	199	34,200
Ammar Yasir	Member of Taliban political office	August 15, 2021	April 1, 2022	157	80,800
Ahmadullah Muttaqi	Assistant Chief of Staff to the Prime Minister, Deputy Director General Public and Strategic Affairs Office	August 15, 2021	March 26, 2022	1281	71,200
Qari Saeed Khosty	Ministry of Information Public Relations and Press Director	August 15, 2021	March 26, 2022	296	190,300
Abdul Wah d Aryan	Director of Bakhtar News Agency	November 16, 2021	April 3, 2022	398	6,587
Anas Haqcani	Senior Taliban member	August 18, 2021	April 15, 2022	301	272,000
Ahmadulla n Wasiq	Radio Television Afghanistan (RTA) Director	August 15, 2021	April 2, 2022	1,099	167,800
Inamullah Samangani	Deputy Spokesperson for the Islamic Emirate of Afghanistan (IEA)	August 15, 2021	April 1, 2022	1,149	62,900
Bilal Karimi	Deputy Spokesperson for the IEA	August 15, 2021	March 31, 2022	777	116,400
Mohammed Naeem Wardak	Spokesman of the Political Office	August 15, 2021	April 4, 2022	1,151	344,900
Muhammed Jalal	No official title	November 4, 2021	March 26, 2022	2,994	78,500
Abdul Qahar Balkhi	MoFA spokesperson	August 22, 2021	March 30, 2022	1,112	106,000
Suhail Shaheen	Taliban Permanent Representative Nominee to United Nations (UN) and Head of Political Office, former Negotiations Team's Member	August 15, 2021	March 25, 2022	467	595,000
Zabihullah Mujahid	Deputy Minister of Culture and Information and Spokesperson	August 15, 2021	March 31, 2022	1,150	527,900
Al Emarah English	Taliban English Twitter account	August 15, 2021	April 6, 2022	1,734	7,432
Total				15,157	

5

TABLE 1.2

Tweets by Language in Final Dataset

Language	Count	Percentage
Pashto	6,426	42.4
Dari/Farsi	3,276	21.6
English	4,662	30.8
Arabic	793	5.2
Total	15,157	100.0

In reviewing the tweets, our team categorized the content as pertaining to international relations, the economy, organization of the Taliban government, treatment of women and girls, connections to militant groups, poppy cultivation and drug flows, and suicide bombers and martyrs (Table 1.3). These categories were selected based on the study scope and because of the high U.S. policy interest in these areas. Since some tweets were categorized as pertaining to more than one theme, the total number of tweets in Table 1.3 (17,223) exceeds the total number of tweets in Table 1.2 (15,157).

Besides Twitter posts, supplemental data were obtained from press conferences, interviews of the Taliban members by local and foreign news outlets, speeches and decrees of Taliban leaders, and cabinet meeting synopses published between August 15, 2021, and March 2022. This subset of the data, collected from different sources over several months, contains videos and texts mainly in Pashto and Dari/Farsi languages. There is some overlap, though not significant, between this subset of data and Twitter posts, as the Taliban spokespersons usually

TABLE 1.3

Thematic Tweet Categories

Tweet Category	Count	Percentage
Other	9,284	53.9
International relations	4,652	27.0
Economy	2,066	12.0
Organization of the Taliban government	440	2.6
Treatment of women and girls	339	2.0
Connections to regional militant groups	301	1.7
Poppy cultivation and drug flows	84	0.5
Suicide bombers and martyrs	57	0.3
Total	17,223[a]	100.0

[a] This figure is bigger than the actual total number of tweets because the tweets related to economics overlap with the "other" and "international relations" categories.

post the highlights of some interviews and speeches on Twitter. We used this complementary dataset as a validation reference and to compare posted content for any particular event.

Structure of the Study

The structure of the analysis presented in the remainder of this report was developed after the tweets were aggregated and coded by individual, language, and theme as we have already described. In parallel to the analysis of tweets, we took a deeper look at the placement and influence of the individuals whose tweets we selected for aggregation and analysis. We judged that understanding the leadership structure was important to better understanding the potential influence and impact of their communications. Chapter 2 provides a brief review of the organization of the Taliban leadership. This will help the reader understand the placement of key spokesmen in the Taliban government and the relative importance of their messaging.

Once tweets were coded by language and theme, we looked for correlations between the themes across the language groups and found a relatively strong relationship between major thematic issues among the languages. The next few chapters of the report examine the themes and the tweets relevant to them. Chapter 3 focuses on economic messaging, Chapter 4 on external relations and connections to regional militant groups, and Chapter 5 on women's rights and education. Chapter 6 provides observations and conclusions derived from our integration of the analysis and some recommendations for further study and analysis.

CHAPTER 2

Taliban Leadership Overview

To better understand Taliban leadership messaging on Twitter, it is important to understand their leadership composition. Recognizing the placement and influence of key leaders and spokespersons helps explain the importance of comments they make on Twitter and other social media. Without this knowledge, policymakers might miss or misinterpret important policy signals or insights into internal Taliban policy differences. We undertook this analysis to identify the most-important Taliban Twitter accounts to enable better analysis of the significance of information being communicated. Although our analysis of the principal policy issues (described in the remaining chapters of this report) focused on what we could learn from the selected Twitter accounts, we supplemented the information obtained from tweets with other media sources to provide additional context for understanding the Taliban leadership and the relative significance of their media communications.

Organization of the Taliban Leadership

The Taliban acting government leadership composition has remained in the same hands, in some cases passing from father to son, since the group's inception in the 1990s. Those in the top echelon are exclusively male, Pashtun, graduates of Pakistan seminaries, UN-sanctioned, and (to some extent) former Guantanamo detainees. After the deaths of Mullah Muhammad Omar, the founder of the Taliban, and Jalaluddin Haqqani, founder of the Haqqani Network,[1] their sons emerged as new leaders of the group. However, the Haqqani Network is more influential and powerful in the group's leadership structure than it was in the 1990s. In their messaging on Twitter, the Taliban consistently talk about the group's cohesion, leadership structure, and hierarchy. This unity is demonstrated both by mutual support to each group's policies and through media coverage of events showing their leaders working together. The Taliban also show in their messaging that topics are divided among the senior leaders of the group, and leaders tend to stay on topic and in their own lanes.

---

[1] The Haqqani Network is a Sunni Islamist militant organization responsible for some of the highest-profile attacks of the Afghan war. The U.S. government designated the Haqqani Network as a foreign terrorist organization in 2012 because of its involvement in the Afghan insurgency and its ties to the Taliban (National Counterterrorism Center, "Haqqani Network," webpage, undated).

Since the emergence of the Taliban, power within the group has been concentrated in the office of their leader, in the military and intelligence wings of the group, in the Finance Ministry, MoFA, and the Ministry of Propagation of Virtue and the Prevention of Vice, and judiciary sector. The individuals leading these ministries are among the most-powerful people within the group hierarchy and leadership.

It is debatable who the most-powerful figures in the Taliban regime are. To assess the power of individuals in the top echelon, we identified five criteria to estimate their authority and influence: (1) control over intelligence, (2) control over the military, (3) control over significant financial resources, (4) control over significant human resources, and (5) ability to set policy priorities. Individuals meeting two of these five criteria are considered to be powerful within the Taliban government (Table 2.1). Since no data are available for quantifying each metric, we use the position in the acting government as a proxy to evaluate the relative power of each individual. Certain individuals—such as Abdul Ghani Baradar, the Acting First Deputy Prime Minister for Economic Affairs—are not included in this list because they do not meet the minimum criteria for this list.

TABLE 2.1
Assessment of Taliban Leadership Influence

Name	Control over Intelligence	Control over the Military	Control over Significant Financial Resources	Control over Significant Human Resources	Ability to Set Policy Priorities
Hebatullah Akhundzadah	Yes	Yes	Yes	Yes	Yes
Sarajuddin Haqqani	Yes	Yes	Yes	Yes	Yes
Haji Mali Khan	No	Yes	No	Yes	No
Mullah Tajmir Jawad	Yes	No	Yes	Yes	No
Abdul Hakim Haqqani	No	No	No	Yes	Yes
Mullah Yaqoob Mujahid	No	Yes	Yes	Yes	No
Abdul Haq Wasiq	Yes	No	Yes	Yes	No
Mullah Hidayatullah Badri	No	No	Yes	Yes	No
Mawlawi Amir Khan Muttaqi	No	No	No	Yes	Yes
Sheikh Mohammad Khalid Hanafi	No	No	No	Yes	Yes
Mullah Fazel Mazloom	No	Yes	No	Yes	No

Senior Taliban Leaders

Hebatullah Akhundzadah

Hebatullah Akhundzadah, or Amir al-Mu'minin (commander of faithful), is the reclusive and powerful leader of the Taliban. Before becoming the leader of the group, he was the Taliban's top judge and deputy of Mullah Akhtar Mansour, ex-leader of the Taliban, who was killed in a drone strike in May 2016.[2] Believed to be in his early sixties, Mullah Hebatullah grew up in the Panjwai district of Kandahar. He is from the Noorzai tribe and comes from a clergy family. His father, Muhammad Akhund, was imam of a mosque in the Safi Rawan village of Panjwai and did not participate in political activities nor seek membership in tribal councils or participate in their disputes.[3] After the Soviet invasion, Mullah Hebatullah and his father moved to Quetta, where Mullah Hebatullah continued his religious studies in one of the first established seminaries in the area.[4] He started his career in the Taliban in the directorate of Vice and Virtue of Farah province and was soon promoted to be an instructor in a Jihadi Madrasa in Kandahar that Mullah Omar personally oversaw.

Mullah Hebatullah is the spiritual leader of the Taliban. On his first visit to Helmand as the de facto ruler of the Taliban, he reportedly visited the graveyard of suicide bombers and paid tribute to them. He has repeatedly urged the Taliban not to interfere in tasks assigned to others and asks them to resolve differences with respect. He has urged fighters to obey their leaders and act faithfully to sharia law.[5] The Taliban covered Mullah Hebatullah's travel to Herat, Helmand, Farah, and Zabul, but little is known about his travel otherwise.[6] He has remained out of the public eye and has not been photographed or filmed since the Taliban seizure of power last August. Taliban spokespersons are very sensitive to publishing any identifiable information on him, including his photograph, traveling schedule, and video clips.

Judging only from tweets by leading Taliban figures and the Taliban media team, it appears that Mullah Hebatullah has three main responsibilities: setting policy priorities and maintaining the group's cohesion, appointing government officials, and guiding the Taliban spiritually. As stated in the Taliban Twitter accounts, the head of the Supreme Court and its members, some cabinet members, deputy ministers, and governors are apparently appointed

---

[2] "Afghan Taliban Announce Successor to Mullah Mansour," BBC News, May 25, 2016.

[3] "Hibatullah's Roots Were Non-Political and Reclusive," TOLOnews, May 2016.

[4] Mujib Mashal and Taimoor Shah, "Taliban's New Leader, More Scholar Than Fighter, Is Slow to Impose Himself," *New York Times*, July 11, 2016.

[5] Arg Presidential Palace [@ARG_1880] (official Twitter account), "Amir-ul-Momineen Sheikh Maulvi Hibatullah Akhundzada's instructions to the directors of public institutions," Twitter post, December 18, 2021.

[6] Muhammad Jalal [@MJalal313] (no official title), "The Amir of IEA, H.E Sheikh Hibatullah Akhundzada paid a visit to Herat province and gave necessary instructions to the officials there. This is the third province that he has visited. Previously he went to Farah province," Twitter post, January 12, 2022.

by him.[7] However, there is little or no knowledge of who appoints powerful ministers, such as those of interior, defense, intelligence, and finance.

In some cases, Mullah Hebatullah's policy guidance is not reflected in action. For example, Taliban spokespersons consistently remind their fighters to respect decisions made by their leaders.[8] Mullah Hebatullah has adopted a public stance of promoting amnesty, likely with the aim of securing support from both the international community and a broader swath of Afghanistan's population. However, this outward commitment to amnesty may not align with the private agenda of his administration, which evidence suggests to be focused on purging former government security officials. For instance, an investigative report by a group of multinational journalists discusses the death of more than 500 former government officials.[9]

Another example of discontinuity between Mullah Hebatullah's policy announcements and implementation is girls' education and women's participation in the workforce—issues frequently discussed in the Taliban's top echelon. For several months, the Taliban promised the world that they would allow girls to receive education. However, that promise was broken after a three-day cabinet meeting chaired by Mullah Hebatullah in March 2022, where it was decided not to allow secondary education for girls.[10] This meeting was attended by all senior Taliban leaders, including Sarajuddin Haqqani. Chapter 5 provides a more extensive discussion of women's rights and education.

Sarajuddin Haqqani

Sarajuddin Haqqani is the second most powerful person in the Taliban regime. He is designated as a global terrorist by the FBI, which offers a $10 million bounty for information leading to his arrest.[11] He is the deputy leader of the Taliban and the Acting Interior Minister.

---

[7] Ahmadullah Muttaqi [@Ahmadmuttaqi01] (Assistant Chief of Staff to the Prime Minister, Deputy Director General Public and Strategic Affairs Office), "By order of Amir al-mu'minin, Sheikh Hadith Maulvi Abdul Hakim Haqqani appointed as the head of Supreme Court, Sheikh Maulvi Mohammad Qasim Turkman as the first deputy, and Sheikh Maulvi Abdul Malik appointed as the second deputy," Twitter post, October 15, 2021.

[8] Ahmadullah Muttaqi [@Ahmadmuttaqi01] (Assistant Chief of Staff to the Prime Minister, Deputy Director General Public and Strategic Affairs Office), "The orders of Commander of the Faithful should be implemented 100 percent," Twitter post, March 23, 2022.

Arg Presidential Palace [@ARG_1880] (official Twitter account), "Amir-ul-Momineen Sheikh Maulvi Hibatullah Akhundzada's instructions to the directors of public institutions," Twitter post, December 18, 2021.

[9] Barbara Marcolini, Sanjar Sohail, and Alexander Stockton, "The Taliban Promised Them Amnesty. Then They Executed Them," *New York Times*, April 12, 2022.

[10] Inamullah Samangani [@HabibiSamangani] (Deputy Spokesperson for the IEA), "The cabinet meeting of the Islamic Emirate was chaired by the Amir al-Mu'minin Hafezullah. On the 16th of Sha'ban 1443 AH, the cabinet meeting of the Islamic Emirate was held in Kandahar province, for three days, chaired the Amir al-Mu'minin," Twitter post, March 23, 2022.

[11] Federal Bureau of Investigation, "Most Wanted: Sirajuddin Haqqani," webpage, undated.

Sarajuddin Haqqani is entrusted with decisionmaking power within the group and is the public face of the leadership in the absence of Mullah Hebatullah. Taliban members usually call him *caliph* or *Khalifah*. Theologically, the caliph is the highest civilian, military, and religious leader of the Islamic State or the successor of the prophet Mohammad—which theologically has more political significance than the title of Amir al-Mu'minin (commander of faithful). Some Taliban members also call him the "defender of Islamic religion" on Twitter.[12]

Messaging about Sarajuddin Haqqani's leadership is posted in Pashto, Dari/Farsi, Arabic, and English. He is portrayed as an inspirational leader who maintains security and meets with foreign delegations. Posts emphasize that there is no such thing as the Haqqani Network or division among the Taliban. In Arabic and Pashto messaging, his leading role in defeating Americans and North Atlantic Treaty Organization (NATO) forces is emphasized repeatedly.[13] In English, he is portrayed as a national leader, a different portrayal than is generally found in the Western media.[14]

Sarajuddin Haqqani transcends any other individual within the Taliban group except Mullah Hebatullah. He appoints ministers, provincial governors, and diplomats. His brother Anas Haqqani represents him in meetings with the international community. Even though Mullah Yaqoob Mujahid, son of Mullah Omar, leads the Defense Ministry, Sarajuddin Haqqani installed his father-in-law, Haji Mali Khan, as Deputy Chief of Staff of the Taliban Defense Ministry.

Sarajuddin Haqqani is the second of seven sons of Mawlawi Jalaluddin Haqqani. As the family's patriarch, Sarajuddin Haqqani leads the clan-based organization that the United States and its allies refer to as the Haqqani Network. Mawlawi Jalaluddin Haqqani was the founder of the Haqqani Network and was a prominent anti-Soviet commander who led operations against the former Soviet Union. Three of Sarajuddin Haqqani's brothers—Badruddin Haqqani, Omar Haqqani, and Mohammed Haqqani—were killed by U.S. forces, and another, Nasiruddin, was assassinated in Islamabad, Pakistan. It is believed that Sarajuddin Haqqani

---

[12] Muhammad Jalal [@MJalal313] (no official title), "Siraj-ud-Din Haqqani, the true defender of Islamic religion," Twitter post, December 31, 2021.

[13] Muhammad Jalal [@MJalal313] (no official title), "Deputy Islamic Emirate and Minister of Interior His Excellency Khalifa Sirajuddin Haqqani, may God protect him. He who fought against the so-called superpower of our time and brought them to their knees in Afghanistan," Twitter post, November 8, 2021.

Anas Haqqani [@AnasHaqqani313] (senior member of the Taliban), "Khalifa Sirajuddin Haqqani, the legendary leader, and one of the great leaders against the invading forces and the leader of the battle of freedom. He gave a new life to his heroism by appearing in front of the journalists' lenses. The defeat of NATO was the result of the loyalty, altruism and leadership of these leaders to the liberation battle," Twitter post, March 5, 2022.

[14] Muhammad Jalal [@MJalal313] (no official title), "The message of H.E Khalifa Sirajuddin Haqqani was different than the way his image was portrayed in western media. He talked about respecting the sovereignty of our country, relations with international community. He talked about the safety & security of Afghans and Afghanistan," Twitter post, March 5, 2022.

was born between 1970 and 1978 in the tribal areas of Pakistan and Afghanistan.[15] He received his early education from his father, then attended Anjuman Uloom Al-Qur'an (a religious seminary in northwestern Pakistan), and then Dar-ul-Uloom Haqqania (a prestigious religious seminary in the eyes of the Taliban, in Akora Khattak near Peshawar).[16]

The Haqqani Network was behind some of the deadliest attacks against U.S. and Afghan military forces and against civilians. The network relied heavily on suicide bombers to attack their targets. In 2009, a suicide bomber attacked a CIA outpost in southeast Afghanistan, killing seven American intelligence personnel.[17] Recently, in an audio communication, Sarajuddin Haqqani claimed responsibility for the suicide attack on the Intercontinental Hotel in Kabul in 2018, which resulted in the death of several civilians, including seven Ukrainians. The United States designated Sarajuddin Haqqani as a global terrorist and officials believe he maintains close ties to al Qaeda.[18]

Haji Mali Khan

Haji Mali Khan, the Deputy Chief of Staff of the Taliban Defense Ministry, was captured by U.S. forces in 2011 in Paktia. He was responsible for military coordination and logistics of the Haqqani Network.[19] In 2019, he was released from prison in a prisoner swap.

Mullah Tajmir Jawad

Mullah Tajmir Jawad serves as the deputy head of the General Directorate of Intelligence. Mullah Tajmir Jawad is a commander of the Haqqani Network and, prior to his current position, ran a suicide bombers' training camp.[20] Mullah Tajmir Jawad is accused of leading and managing high-profile suicide attacks and is entrusted with the most-sensitive issues within the Haqqani Network. Furthermore, Khalil-ur-Rahman Haqqani, the uncle of Sarajuddin Haqqani, is the acting Minister of Refugees.

Abdul Hakim Haqqani

Abdul Hakim Haqqani is the acting Chief Justice of the Taliban regime. He belongs to the Ishaqzai tribe, was born in 1967 in Panjwai District, Kandahar Province, and graduated from

---

[15] Rewards for Justice, "Sirajuddin Haqqani," webpage, undated.

[16] Arshad Yusufzai, "Sirajuddin Haqqani, Feared and Secretive Taliban Figure, Reveals Face in Rare Public Appearance," *Arab News PK*, March 7, 2022.

[17] Yochi Dreazen, "The Taliban's New Number 2 Is a 'Mix of Tony Soprano and Che Guevara,'" *Foreign Policy*, July 31, 2015.

[18] Federal Bureau of Investigation, undated.

[19] Michael Semple, "The Capture of Mali Khan," *Foreign Policy*, October 10, 2011.

[20] Abubakar Siddique and Abdul Hai Kakar, "Al-Qaeda Could Flourish with New Strategy Under Taliban Rule," Radio Free Europe/Radio Liberty, September 30, 2021.

Haqqani Seminary of Pakistan.[21] His views are key to understanding Taliban policies on women's participation in the workforce and education. In his recent book, *The Islamic Emirate and Its System* (with a preface written by Mullah Hebatullah), Abdul Hakim Haqqani extensively discusses his views on women's participation in the workforce and education. He says that women are deficient in intellect (compared with men), should stay at home, must not mix with men, and that those who entrust women with public positions "will not be saved" in the life after death.[22] According to him, "women should not be involved in politics. Their main function is to bear and raise children."[23] He says that women "even do not have the right to go to mosques to worship God" and that "if women are not allowed to pray in the mosques, then shopping in market and visiting or working in the government offices is clear."[24]

Abdul Hakim Haqqani believes that women can study only a few topics—not, for example, science, engineering, or technology. Studies preferably should be administered at home by male guardians. "It should be said that women can choose fields worthy of their dignities, such as medicine and tailoring. However, there is no need for women in chemistry and geometry and the like. Men alone meet the needs of society in all these fields."[25]

Mullah Yaqoob Mujahid

Mullah Yaqoob Mujahid is the Taliban's acting Minister of Defense. He enjoys significant symbolic respect within the Taliban as the eldest son of Mullah Omar, the founder and late supreme leader of the Taliban. Mullah Yaqoob was born in 1990 and is a graduate of several seminaries in Karachi, Pakistan.[26]

In the Taliban messaging on Twitter, Mullah Yaqoob focuses on Afghanistan's borders and the construction of an army for the Taliban. He regularly pays visits to the borders in the south, west, and north of Afghanistan and tries to project the Taliban's power through his visits with the public and to neighboring countries.[27] Recently, after an exchange of fire on the border of Afghanistan and Iran between the Taliban and Iranian security forces and other border issues between the two groups, Iranian officials met with Mullah Yaqoob to resolve the challenges.[28] The Taliban widely shared Mullah Yaqoob's threatening language

---

[21] Afghan Biographies, "Ishaqzai, Abdul Hakim Mawlawi Sheikh," webpage, July 13, 2022.

[22] Abdul Hakim Haqqani, *The Islamic Emirate and Its System*, Darul-Ulum Shariyah, 2022, p. 80.

[23] Abdul Hakim Haqqani, 2022, p. 151.

[24] Abdul Hakim Haqqani, 2022, p. 151.

[25] Abdul Hakim Haqqani, 2022, p. 262.

[26] Frud Bezhan, "The Rise of Mullah Yaqoob, the Taliban's New Military Chief." Radio Free Europe/Radio Liberty, August 27, 2020.

[27] Muhammad Jalal [@MJalal313] (no official title), "Picture is from Defense Minister's visit to the southern and western borders of the country," Twitter post, March 17, 2022.

[28] Ministry of National Defense, Afghanistan [@modafghanistan2] (official Twitter account), "A delegation from the Ministry of Defense met with the Deputy Chief of Mission of the Islamic Republic of Iran in

against Pakistan on the anniversary of his father's death, which he seized as an opportunity to threaten retaliation for any future military attack within Afghanistan's borders.[29] (At the time, Pakistani forces had conducted military operations inside Afghanistan.)

Abdul Haq Wasiq

One of the most powerful organizations within the Taliban group is the General Directorate of Intelligence. The organization is led by Abdul Haq Wasiq, a former Guantanamo detainee. The mission of the General Directorate of Intelligence is to protect "national interests" and "Islamic values," as stated by the Twitter account of the agency.[30] Tweets of the agency claim that it is mainly focused on fighting the Islamic State Khorasan Province (ISKP), finding kidnappers, and locating and confiscating caches of weapons.

Wasiq was born between 1971 and 1975 in Gharib village of Khogyani District in Ghazni Province.[31] He began his early religious education under his father's supervision. After his father's death, he continued his education in Pakistan seminaries with the financial help of his brothers. For a while, he led prayers, as an imam, in a district of Ghazni; however, he had affection and alliance with Gulbuddin Hekmatyar, the leader of Hezb-e-Islami.[32] After the Taliban took control of Afghanistan, he moved to Kabul and was assigned as the Deputy of the General Directorate of Intelligence, reporting to Qari Ahmadullah, the head of that agency. In that position, he managed the relationship with al Qaeda and their training camps in Afghanistan.[33] He was captured in Ghazni, arrived at Guantanamo Bay in 2002, and was exchanged in a prisoner swap in 2014 for SGT Beaudry Robert "Bowe" Bergdahl, a U.S. Army

---

Kabul," Twitter post, April 26, 2022.

[29] "Maulvi Muhammad Yaqub Mujahid's Speech in the 9th Year Celebration of Amirul Momineen Mullah Muhammad Umar Mujahid," Radio Television Afghanistan (RTA) Pashto, 2022.

[30] General Directorate of Intelligence [@GDI1415], "Dear Countrymen! With the divine assistance of almighty Allah (SWT), the unyielding support of the courageous Afghan nation, and the unbelievable sacrifices of our holy warriors, our beloved country has once again been blessed with a truly Islamic government, and security that is steadily prevailing throughout the country. The leadership and staff at the General Directorate of Intelligence of the Islamic Emirate of Afghanistan consider protection of national interests and Islamic values as their religious duty and national obligation, and they proudly endure any tiresome of hardships to ensure the security of the countrymen. The General Directorate of Intelligence urges all the compatriots to assist us in establishing absolute security and foiling the nefarious plots of evil elements in a timely manner by urgently reporting any suspicious activity from individuals within your regions to local security forces, or to call us free toll number 1001. Regards, the General Directorate of Intelligence (GDI) 26/10/2021 Gregorian," Twitter post, October 26, 2021.

[31] *Official Journal of the European Union*, "2014/142/CFSP of 14 March 2014 implementing Decision 2011/486/CFSP concerning restrictive measures directed against certain individuals, groups, undertakings and entities in view of the situation in Afghanistan," Vol. 57, March 15, 2014, pp. 6–8.

[32] *Official Journal of the European Union*, 2014, pp. 6–8.

[33] *Official Journal of the European Union*, 2014, pp. 6–8.

soldier who was held captive from 2009 to 2014 by the Haqqani Network.[34] Wasiq is known for his brutal interrogation methods against the Taliban's opponents.[35]

Mullah Hidayatullah Badri

Mullah Hidayatullah Badri is the Minister of Finance in the Taliban regime and an UN-sanctioned member of the Taliban. The Ministry of Finance is the most powerful financial organization that plays a key role in financing the Taliban regime. Mullah Hidayatullah is focused on generating revenues through taxation and customs. Increasing revenues is a constant focus of the group's spokespersons on Twitter.

Mullah Hidayatullah was born around 1972 in Maiwand District of Kandahar Province and belongs to the Ishaqzai tribe. He was the childhood friend of Mullah Omar and served as one of his closest advisers and as his chief financial officer.[36] The two were so close that Mullah Hidayatulla was living in the presidential palace with Mullah Omar and, at one point, was responsible for deciding who could meet with the group leader. From 2001 until 2018, Mullah Hidayatulla led the financial commission of the Taliban and was financing suicide bombers in Kandahar province.[37] He is on the sanction list of the UN Security Council for "being associated with Al-Qaida, Usama bin Laden or the Taliban."[38]

Mawlawi Amir Khan Muttaqi

Mawlawi Amir Khan Muttaqi is the UN-sanctioned acting Foreign Minister of the Taliban. He is the face of the Taliban's diplomacy with the Western world, Arab countries, China, and Russia. His visits to world capitals and his meetings with foreign officials are covered extensively by the Taliban on Twitter.

Muttaqi's ancestors are from Paktia; he was born on February 26, 1971, in the Nad Ali district of Helmand.[39] After the invasion by the Soviet Union, he moved to Pakistan and attended seminary schools in refugee camps. He joined the Taliban in early 1994, and after the fall of Kandahar that year, became a member of the Taliban High Council and head of Kandahar Radio.[40] When Kabul fell under the control of the Taliban, he became the Acting Minister of Information and Culture in 1996. In 1997, he was appointed as the Chief of Staff of MoFA; three years later, in 2000, became the Minister of Education until the fall of the

---

[34] "Taliban's Mullah Omar Celebrates Prisoner-Swap 'Victory,'" BBC News, June 1, 2014.

[35] *Official Journal of the European Union*, 2014, pp. 6–8.

[36] *Official Journal of the European Union*, 2014, pp. 6–8.

[37] *Official Journal of the European Union*, 2014, pp. 6–8.

[38] *Official Journal of the European Union*, 2014, pp. 6–8.

[39] MoFA, "Biography Minister of Foreign Affairs of Afghanistan," webpage, undated.

[40] MoFA, undated.

Taliban regime. In 1999, he was the chief negotiator for the Taliban with the Northern Alliances. After the fall of the Taliban regime, he led the Cultural Commission of the Taliban.[41]

Sheikh Mohammad Khalid Hanafi

Little is known about Sheikh Mohammad Khalid Hanafi. He is a religious scholar and holds extreme views on women and the implementation of sharia law. He leads the Ministry of Propagation of Virtue and the Prevention of Vice. His organization affects every aspect of men's and women's public and private lives in Afghanistan. In his first move, he took over the building of the Ministry of Women's Affairs of the former republic.

The Ministry of Propagation of Virtue and the Prevention of Vice oversees monitoring and implementing sharia law in every department of the Taliban group, every administrative unit of Afghanistan, and every individual in Afghanistan. The ministry has already forced university students to enroll in religious subjects. It forced government employees to wear a hat and prohibited the use of neckties.[42]

Mullah Fazel Mazloom

Mullah Fazel Mazloom is the First Deputy Minister of Defense in the Taliban regime. He is treated as one of the most-respected field commanders among the Taliban and is a role model on Twitter for their fighters. Mullah Fazel is a former Guantanamo detainee brought to the prison the day it opened and eventually also part of the 2014 swap for Bergdahl.[43] Tweets of the Taliban show that he is mainly in charge of leading and managing operations against the National Resistance Forces in the north of Afghanistan.

Mullah Fazel belongs to the Kakar tribe and was born between 1963 and 1968 in Uruzgan province.[44] He was another close associate of Mullah Omar, spent time at the Al-Farouq camp established by al Qaeda, assisted the Islamic Movement of Uzbekistan, and is accused of being responsible for killing thousands of persons of the Shia minority.[45]

---

[41] MoFA, undated.

[42] "An Overview of the Activities of Violation of Citizens' Rights by the Ministry of Propagation of Virtue and the Prevention of Vice," *Etilaatroz Daily*, May 14, 2022.

[43] "The Guantánamo Docket," *New York Times*, May 18, 2021.

[44] *Official Journal of the European Union*, 2014, pp. 6–8.

[45] *Official Journal of the European Union*, 2014, pp. 6–8.

CHAPTER 3

The Economy

The political regime changes on August 15, 2021, entailed significant economic disruption in Afghanistan, a country with a fragile economy and heavily dependent on foreign aid. This political crisis disrupted the banking system, domestic and foreign trade, and international money transfer, and it led to a sizable depreciation of Afghan currency and the suspension of international development assistance. As a result, unemployment rose and millions of Afghans now face acute food shortages.[1] Based on some projections, by the end of 2022, the real gross domestic product per capita will decline by around 30 percent.[2] The Taliban messaging on Twitter and the local press suggests that they acknowledge the economic hardships, but they disagree with the bleak picture painted by the World Bank and UN agencies. Instead, the Taliban messaging to Afghans is that the ongoing economic crisis is temporary. The Taliban are attempting to formulate policies to improve the economy and attain self-reliance by investing in agriculture and the extractive industry and by supporting and developing regional initiatives, including regional pipeline construction. The Taliban's messaging to the international community is that Afghanistan needs international support for development projects.

Along with assurances addressing economic hardship, the Taliban are attempting to craft a narrative to dodge responsibility for the 2022 economic crisis. To deflect attention from the reality of the economic situation, they claim the economy was already crumbling because of rampant corruption under the previous regime. They also claim that the U.S. blockage of Afghan currency reserves is aggravating the situation. In addition, some Taliban officials addressing gatherings in the provinces have stated that the U.S. presence in the past 20 years damaged the country and the economy, so the regime expects the United States to help rebuild Afghanistan. In their general messaging, Taliban members use strong language while addressing the public about the U.S. presence in Afghanistan, employing such phrases as "occupation," "Satanic forces,"[3] and "aggressors"—but they markedly temper their tones for nonlocal audiences.[4]

---

[1] World Food Programme, "Afghanistan Emergency," webpage, undated.

[2] World Bank, "Urgent Action Required to Stabilize Afghanistan's Economy," April 13, 2022.

[3] RTA, "Sher Mohammad Abbas Stanikzai: We Have Always Turned Those Blind Who Had Bad Intent About Afghanistan," video, December 29, 2021.

[4] Information TV, "Sher Mohammad Abbas Stanikzai' Speech in the West of Kabul," video, December 29, 2021.

Taliban officials have repeatedly stated that the economy is the top priority for the government. They have adopted an economic-centered foreign policy, particularly toward the neighboring and regional countries. The guidelines and decrees issued by the putative Taliban supreme leader (Amir al-Mu'minin) are not explicit about addressing the economic issues. However, Prime Minister Mullah Hassan Akhund has made remarks about the economic situation and issued guidelines. In one instance, referring to the economic hardship, he said, "it is not incumbent upon Taliban to feed Afghans, but God will do that."[5] He outlined the Taliban's economic policies more formally in his speech at the first economic conference held in mid-January 2022 in Kabul. Representatives of 60 countries, UN agencies, and private-sector actors attended this conference.[6] Akhund said that the Taliban would improve the banking system, focus on agriculture and mining, and expect the international community to provide development assistance without imposing stringent conditions to preclude economic dependency. He asserted that the security situation is stable and that corruption has been eradicated, thus creating an environment that bolsters economic growth and creates a fair economic system. Other Taliban officials echoed a similar idea at this conference, adding that improving the tax system and revenue-collection mechanism and strengthening the private sector are integral elements of their economic policies. The key communication to the international community at this conference was that Afghanistan needs development assistance, but the distribution of that aid needs to change because aid provision for the past 20 years was inefficient and resources were embezzled. The Taliban are conveying similar messaging to the Afghan public about corruption and aid efficiency, but they reverse the emphasis, downplaying the significance of the foreign aid and highlighting that the country is endowed with ample natural resources that serve as the engine of economic growth.

Organizational Context and Messaging Channels

The synopses of the Taliban's cabinet meeting reports posted on Twitter suggest that topics related to economics—such as taxation, revenue management, banning the use of foreign currencies in the local market, banking system, and import and export of goods—are on the agenda for every meeting. The prime minister chairs these meetings, but most of the day-to-day economic affairs at the operational level are handled by the Economic Commission, chaired by First Deputy Prime Minister Mullah Abdul Ghani Barader.[7] Taliban Twitter

---

[5] Ahmadullah Wasiq [@WasiqAhmadullah] (RTA Director), "You can listen the live speech of Mullah Hassan Akhund," Twitter audio post, November 27, 2021.

[6] Inamullah Samangani [@HabibiSamangani] (Deputy Spokesperson for the IEA), "In today's economic conference, the representatives of 20 countries attended in person and the representative of other 40 countries virtually. The participants will discuss the economic issues of Afghanistan, particularly the banking problem and challenges facing the private sector, and recommend solutions," Twitter post, January 19, 2022.

[7] These documents are only in the local languages (Pashto and Dari/Farsi). The Ministry of Finance, the central bank, the Ministry of Economy, and the Ministry of Industry and Commerce are members of this commission.

posts on economic affairs are mainly in the local languages, and it is evident that their target audiences are primarily Afghans. In addition to Twitter, the Taliban broadcast their economic messaging through the state news outlets, Bakhtar News Agency, and the RTA. The three leading spokespersons of the de facto government and the spokespersons for the central bank, the Ministry of Finance, and the Ministry of Economics, regularly appear on local media and discuss economic policies. It is worth noting that the Taliban have not yet presented any official documents outlining their development strategy and economic policies.

Fiscal and Monetary Policies

The Taliban claim that Afghanistan has, for the first time in several decades, been able to prepare the national budget without being dependent on foreign aid. They circulated this claim widely on Twitter and local media. This unprecedented progress has been attributed to the fight against corruption and improved transparency in revenue collection, particularly in the customs and extractive industry. They point out some examples that indicate that, despite economic anomalies, the Taliban have collected more revenue from specific sectors in each period compared with previous regimes. Taliban authorities, such as the Ministry of Mining and Petroleum and the Ministry of Finance, reported some figures indicating increased revenue.[8] The Taliban have also stated that although such expenses as civil servants' salaries and other operational costs constitute a significant portion of the national budget for fiscal year 2022–2023, resources have also been earmarked for some development projects. Taliban officials have promised to finance incomplete infrastructural projects, such as dams, irrigation canals, roads, and a new energy project—the Turkmenistan-Afghanistan-Pakistan-India (TAPI) pipeline—out of national revenues. The key concept that the Taliban are trying to convey to the public is that, unlike in the previous regime, national revenues are not embezzled. The impression the Taliban seek to create is that the administration can put the Afghan economy on a growth trajectory and gain self-sufficiency. However, these rosy pictures are in stark contrast with realities on the ground. For months, Taliban have only sporadically disbursed civil servants' salaries,[9] and hospitals are running out of essential supplies.

The value of Afghan currency has depreciated significantly since the return of the Taliban to power. Political turmoil, flight of capital from Afghanistan, and shortage of the U.S. dollar in the domestic market caused fluctuation in the exchange rate. Despite Taliban assurances before entering Kabul, disruption in the banking system and declining public trust in banks compounded the monetary problem. Taliban officials capped weekly bank withdrawals, banned international transfers from individual and corporate accounts, and froze the accounts of all the former senior government officials to avoid the collapse of the banking system. Taliban officials have repeatedly stated that the banking problem is caused

[8] TOLOnews, "TOLOnews 6pm News—24 March 2022," video, March 24, 2022b.

[9] William Byrd, "Taliban Are Collecting Revenue—But How Are They Spending It?" Institute of Peace, February 2, 2022.

by restricted access to the international monetary system and the freezing of Afghan currency reserves.[10] They welcomed the general licenses issued by the U.S. Department of the Treasury because these licenses relaxed restrictions on the transfer of humanitarian aid and interbank transactions.[11]

The Afghan banking system is experiencing an acute liquidity shortage under the Taliban regime.[12] Funneling the disbursement of humanitarian aid through the banking system is the main source of infusing cash into the system. The financial packages of humanitarian aid are deposited in a private commercial bank, and the disbursement is made in Afghan currency supplied by the central bank, Da Afghanistan Bank.[13] Taliban posts on Twitter provide regular updates on the arrival of humanitarian aid cash packages, which are deposited in a private bank in Kabul.[14]

Promoting the use of Afghan currency in day-to-day transactions, cracking down on smuggling dollars to neighboring countries, and selling more U.S. dollars on the local market by the central bank were measures meant to stabilize monetary fluctuation. Taliban Twitter posts indicate that the banking system has been frequently discussed in their meetings with the delegation of other countries and UN representatives.

The Taliban economic framework hinges on trade and regional connectivity, the extractive industry, agriculture, and small-scale businesses. Taliban messaging on Twitter and local media is informative about the first three areas—trade and regional connectivity, the extractive industry, and agriculture—and we discuss these here.

---

[10] Abdul Qahar Balkhi [@QaharBalkhi] (MoFA Spokesperson), "The freezing of foreign exchange reserves and the associated banking crises deteriorates the humanitarian crisis further," Twitter post, December 21, 2021.

[11] U.S. Department of the Treasury, "Treasury Issues Additional General Licenses and Guidance in Support of Humanitarian Assistance and Other Support to Afghanistan," December 22, 2021.

[12] Michelle Nichols, "U.N. Warns of 'Colossal' Collapse of Afghan Banking System," Reuters, November 22, 2021.

[13] The humanitarian aid money is deposited in the corporate accounts that the aid agencies have with a private commercial bank (the Afghanistan International Bank), then the disbursements are made directly from these bank accounts. This disbursement channel has been defined to circumvent the influence of the Taliban over aid resources.

The Taliban have banned the use of foreign currency in domestic markets. Therefore, the aid money should be converted into *Afghani,* the Afghan currency supplied by the central bank (Da Afghanistan Bank). However, the head of the United Nations Development Programme once pointed out that his program cannot spend US$135 million in humanitarian aid because the central bank failed to convert it into *Afghani* (Michelle Nichols, "U.N. Has Millions in Afghanistan Bank, but Cannot Use It," Reuters, February 3, 2022).

[14] Zabihullah Mujahid [@Zabehulah_M33] (Deputy Minister of Culture and Information and Spokesperson), "Officials at Da Afghanistan Bank say, as part of series of humanitarian aid to #Afghanistan, $32 million in cash arrived in #Kabul today (Monday, February 21) and transferred to the Afghanistan International Bank (AIB)," Twitter post, February 21, 2022.

Trade and Regional Connectivity

Uzbekistan, Turkmenistan, and Kazakhstan are the major regional economic players in Central Asia. Tajikistan, Afghanistan's other neighbor to the north, is not a major economic partner, as is explained in the next chapter on regional political relations. Taliban Twitter posts suggest that Uzbekistan, Turkmenistan, and Kazakhstan have a key role in the Taliban's adopted foreign policy (which is centered on economics, particularly with these neighboring countries that make up their major trading partners). Twitter posts indicate that trade, energy projects, and attracting foreign investment in Afghanistan have been discussed frequently in meetings of Taliban officials with delegations from Turkmenistan, Uzbekistan, Kyrgyzstan, Iran, Pakistan, and China.[15] The Taliban are building on initiatives started under the previous regime and are subscribing to the notion that Afghanistan has the potential to turn into a transit country connecting the Central and South Asian regions. The Taliban also note the potential economic gains from transborder energy projects and transit lines passing through Afghanistan.

Afghanistan is heavily dependent on central Asian countries for energy and for importing other essential commodities. Taliban officials have emphasized the two large energy projects, TAPI and the Turkmenistan-Uzbekistan-Tajikistan-Afghanistan-Pakistan Interconnection Concept, connecting the energy-rich Central Asian countries to the energy-poor South Asian region. The TAPI pipeline has captured significant attention; the Taliban have repeatedly stated that security concerns will not hinder the implementation phase and that they will be able to pay for this project out of the national budget. Transborder trade (and the extension of the railway through Afghanistan, Uzbekistan, and Turkmenistan) has been another area of interest for the Taliban. The railway has been instrumental in the cross-border flow of goods between Afghanistan and its Central Asian neighbors.

Similarly, the Taliban have highlighted the significance of Afghanistan as a transit route for transportation of goods from south to central Asia. The Taliban claim that they have resolved lingering customs issues, expedited the movement of goods, and allocated funding for the expansion of railways to foster trade and regional connectivity. The Taliban send positive posts about the potential economic gains from trade and transit. They emphasize that the stable security situation in Afghanistan is conducive to investment, seeking to encourage Afghan and foreign investors to move funds to Afghanistan and invest in different sectors, particularly the mining industry. Nevertheless, Taliban messaging is ambiguous about funding sources for the implementation of the energy projects and transit lines. Whenever Taliban spokespersons and officials are asked about economic hardship and dwindling development aid, they highlight the role of Afghanistan as a vital transit route that will bolster economic growth and overcome the ongoing economic crises.

---

[15] Based on the word frequency of country names in the Taliban tweets with an economic focus, China tops the list at 428, followed by Pakistan (262), Turkmenistan (239), Iran (160), Uzbekistan (132), and Russia (54). From this, one can infer that the Taliban had more interaction with China than other countries.

Extractive Industry

Mining has become a central element of Taliban economic policies. Taliban messaging highlights that the extractive industry will serve as the lifeblood of the Afghan economy. They repeatedly mention that Afghanistan is endowed with abundant natural resources and minerals that will turn into the engine of economic growth.[16] This characterization is reflected in posts to the public and Taliban interaction with neighboring and regional countries, particularly China. Taliban Twitter posts suggest that investment in the extractive industry has been a significant topic in meetings with Chinese delegations and, to some extent, with delegations from Iran and Turkey.[17] The Chinese government has expressed interest in expanding its investment in the mining industry of Afghanistan.[18] Similarly, the Taliban have requested that the China Metallurgical Group Corporation (or MCC Group) resume its work on the Aynak copper project according to the terms of the contract signed with the previous regime in Afghanistan.[19] In their posts, the Taliban encourage foreign and domestic companies to invest in the mining sector, emphasizing that there will be no barriers and that security will be provided for the mining sites and workers.

In the past 20 years, the Taliban have consistently targeted the resource-rich parts of Afghanistan, and illegal mining was a significant source of income. Now controlling the entire country, the Taliban have disproportionately focused on the extractive industry once again. The Ministry of Mining and Petroleum is among the very few ministries that have gained significant visibility. Taliban spokespersons regularly tweet that the Ministry is fighting illegal mining and is adamant about curbing corruption in the bidding and contracting process of mining projects.[20] The Ministry has primarily focused on small-scale mining;

---

[16] Inamullah Samangani [@HabibiSamangani] (Deputy Spokesperson for the IEA), "Mullah Abdul Ghani Barader on RTA: We support investment in legal mining and export of minerals to create jobs and attain economic self-sufficiency," Twitter post, January 3, 2022.

[17] Abdul Qahar Balkhi [@QaharBalkhi] (MoFA Spokesperson), "Also the Foreign Minister requested the Chinese Ambassador to expedite the work on mining projects that China has invested in," Twitter post, February 3, 2022.

[18] First Deputy Prime Minister Office [@FDPM_AFG], (official Twitter account of the Office of Mullah Abdul Ghani Barader), "and invest in mining, creating economic zones and energy production. The Chinese Foreign Minister added that they will soon start extracting the Mes Aynak mine," Twitter post, March 24, 2022.

[19] TOLOnews, "TOLOnews 6pm News—14 March 2022," video, March 14, 2022a.

[20] Al Emarah English [@Alemarahenglish] (official Taliban English Twitter account), "Minister of Mines at the AfG Economic Conference: Illegal mining has been stopped across the country. Dozens of small mines are being contracted out to domestic investors for legal extraction. Many large mines that require high capita," Twitter post, January 19, 2022.

these projects include coal, chromite, and precious and decorative stones. Taliban spokespersons post lists of tenders and results of the bidding process.[21]

Despite the claim of transparency, little is known about the winners of the mining contracts. The Kabul chamber of commerce has implicitly expressed concerns about the low participation rate of domestic companies and the steep hike in royalties.[22] Although not confirmed by any independent sources, there are suspicions and allegations that senior Taliban members have vested interests in the mining projects. For instance, the Taliban announced that the revenue from exporting coal to Pakistan has doubled.[23] Later, a source familiar with Taliban senior officials claimed that relatives of some Taliban members are running the coal businesses in Pakistan.[24]

Oil and gas also have the Taliban's attention. On April 7, 2022, Abdul Ghani Barader, Deputy Prime Minister for Economic Affairs, inaugurated the drilling of the Qashaqari oil reserve in the northern part of Afghanistan.

Although it is unclear how much revenue the Taliban collect from the extractive industry and how transparent the contracting process is, messaging to the public is not supported by action. Senior Taliban members constantly claim that the only way to achieve economic prosperity and self-reliance is to invest in and extract the natural resources of Afghanistan. They believe that the mining project will generate a massive amount of revenue invested in infrastructure, agriculture, and other sectors of the economy, but expectations in these investments have not been realized.

Agriculture

In 2020 (the most recent record available), 74 percent of Afghanistan's population lived in rural areas, and farming was the main economic activity for rural households.[25] Since taking power, Taliban messaging on agriculture is exclusively directed at local audiences; these posts describe agriculture as the backbone of the Afghan economy and promise to provide support to the farmers, market agricultural products, modernize farming practices, and build irrigation canals and water dams to deflect the effect of droughts. In 2022, the Taliban celebrated Farmer's Day, and senior Taliban officials attended several exhibitions of agricultural products and sent ambitious communications to the farmers. The Ministry of Agriculture,

---

[21] Inamullah Samangani [@HabibiSamangani] (Deputy Spokesperson for the IEA), "The Ministry of Mining and Petroleum has opened the bidding process for ten small-size mining projects in Kandahar, Logar, and Zabul. The interested entities can submit their application within 15 working days and receive the other required documents after the initial screening," Twitter post, January 23, 2022.

[22] TOLOnews, "TOLOnews 6pm News—06 April 2022," video, April 6, 2022d.

[23] "Taliban Has Doubled Coal Prices," DW, July 7, 2022.

[24] Afghanistan International, "The Taliban Are Playing Double Games," video, April 13, 2022.

[25] World Bank, "World Development Indicators," 2020.

Irrigation, and Livestock—the lead policymaking agency—said that all development projects initiated in the previous administration would continue without disruption.[26] However, the feasibility of these projects is in doubt: Most were funded through foreign aid that has dried up.[27] The Taliban messaging and promises to improve agriculture have not been translated into action yet aside from some initial work on irrigation.

That said, the Taliban have followed in the footsteps of the previous regime and focused on water management. They promised farmers that water would be diverted from reservoirs and the Kamal Khan dam for agricultural use, and that new dams and irrigation canals would be constructed. To this end, the Taliban inaugurated construction of the Qosh-Tipa irrigation canal, claiming it to be the largest of its kind in Afghanistan. This three-phase, 285-km canal stretches across several northern provinces.[28] Several senior Taliban officials attended the inauguration ceremony and highlighted the economic significance of the canal. The Taliban had notified the media of this event in advance, so it garnered significant coverage. The Taliban took credit for the canal initiative and labeled it a historic move toward achieving self-sufficiency in wheat production because the canal will irrigate vast agricultural lands that are barren.[29] Taliban messaging promised canal construction would be completed in six years and asserted that the funding would come from domestic revenues.

Poppy Cultivation and Production

In 2021, poppy was cultivated on 177,000 hectares in Afghanistan, making the country the world's leading poppy producer.[30] Poppy cultivation and the fight against drugs were rarely captured in Taliban messaging immediately after the group regained control of the country. In April 2022, however, the Taliban supreme leader issued a decree banning poppy cultivation and production.[31] The release of this decree was one of the rare occasions when the Taliban

---

[26] TOLOnews, "TOLOnews 6pm News—01 April 2022," video, April 1, 2022c.

[27] Ministry of Agriculture, Irrigation and Livestock, "Programs," undated. The webpage advertises a number of projects, but all were initiated using aid money prior to 2020.

[28] Bakhtar News Agency [@BakhtarNA] (state news agency), "Mullah Abdul Ghani Barather, the First Deputy Prime Minister, traveled to the Balkh province to attend the inauguration ceremony of the Qosh-Tipa canal. Several other senior officials have also traveled for this purpose. This canal has a length of 285 kilometers, width 100 meters, and depth 805 meters," Twitter posts, March 29, 2022.

[29] Karimi, Bilal [@BilalKarimi21], Deputy Spokesperson, "Zabihullah Mujahid: today, we are thrilled to roll out the Qosh-Tipa canal. With the completion of this project, we will not be dependent on grains from other countries, and public support is critical to the success of this project. The Islamic Emirate will use all the possible means at its disposal to complete this project," Twitter post, March 30, 2022.

[30] United Nations Office of Drug and Crime, *Afghanistan Opium Survey 2021—Cultivation and Production*, March 22, 2022.

[31] This decree was issued at a very sensitive time when the Taliban were under scathing criticism by the public and the international community for banning girls from going to school. Before the press conference on this decree, the speculation was that the decree would be about rescinding the ban on girls' education.

invited the media to a press conference at the Ministry of Interior. The Taliban spokesperson announced this press conference in multiple languages on Twitter and the local media. The decree is sternly worded and warns of dire ramifications if violated: "As per the decree of the supreme leader of the Islamic Emirate of Afghanistan (IEA), All Afghans are informed that from now on, cultivation of poppy has been strictly prohibited across the country. If anyone violates the decree, the crop will be destroyed immediately, and the violator will be treated according to Sharia law. In addition, usage, transportation, trade, export, and import of all types of narcotics such as alcohol, heroin, tablet K, hashish, etc., including drug manufacturing factories in Afghanistan are strictly banned."[32] The Taliban messaging claims that opium production was zero in 2001, but after 20 years and spending a tremendous amount of money, Afghanistan tops the list of opium producers. This decree, they claim, will lead to the eradication of poppy cultivation and production in Afghanistan.

The Taliban have been silent about the enforcement of the poppy ban decree, particularly because the unveiling of the decree coincided with the peak of the poppy harvesting season. However, some of their messaging highlights their approach to curbing opium production and its illicit trade. Their messaging to the international community is that the Taliban are determined to fight poppy cultivation, production, and trade. Reassuring communications have also been sent to farmers that the government would create other jobs and provide alternative livelihoods. Judging by this messaging, the Taliban have restarted the counternarcotic prosecution office and court and appointed heads of counternarcotics departments in the province.[33] Similarly, they launched a program to rehabilitate millions of drug addicts who created a swiftly growing domestic market for opium products.[34] The Taliban acknowledge the enormity of the drug addiction problem and seek support to address this issue, but they portray the previous Afghan government and its international partners as the culprits: "Drug addicts are being rounded up and taken to rehabilitation centers. IEA's war on drugs con-

---

[32] Inamullah Samangani [@HabibiSamangani] (Deputy Spokesperson for the IEA), "As per the decree of the supreme leader of Islamic Emirate of Afghanistan (IEA), All Afghans are informed that from now on, cultivation of poppy has been strictly prohibited across the country. If anyone violates the decree, the crop will be destroyed immediately," Twitter post, April 2, 2022.

[33] Supreme Court of Afghanistan [@SupremeCourtAfg] (official Twitter account), "The IEA leadership has approved the establishment of the counter-narcotics court," Twitter post, January 24, 2022.

Muhammad Jalal [@MJalal313] (no official title), "The Ministry of Interior has appointed anti-narcotics officials in all provinces. The Deputy Minister for Counter Narcotics has said that he has made recommendations to officials to stop the sale & purchase of narcotics & poppy cultivation in all provinces as soon as possible," Twitter post, December 11, 2021.

[34] Qari Saeed Khosty [@SaeedKhosty] (Ministry of Interior Public Relations and Press Director), "2718 drug addicts taken to hospitals in Kabul Over the past week, at least 2718 drug addicts were collected from various areas in Kabul & were sent to drug treatment centers in Kabul. According to IM Officials, collected addicts were taken to 14OPD cntrs & 14inpatient hospitals," Twitter post, November 20, 2022.

tinues. During the last 19 years of foreign occupation many foreign backed warlords were involved in drug trade and opium cultivation."[35]

Some sporadic efforts notwithstanding, the drug war is a conundrum for the Taliban. Some sources claim opium prices doubled right after the issuance of the decree, with no reaction from the Taliban.[36] Taliban Twitter posts do not address the prosecution of thousands of prisoners accused of smuggling drugs and operating in the orbit of the drug mafia that controls the operation. Similarly, Taliban posts do not respond to allegations that they have inherited a massive reservoir of drugs confiscated by the previous regime. The Taliban likely will sell this contraband to their traditional drug networks in the region and across the globe.[37]

Humanitarian Aid

Taliban messaging on humanitarian aid is twofold. They welcome the humanitarian response and appreciate aid organizations' efforts in collecting and mobilizing more resources. According to Twitter posts, Taliban officials (including the ministers and deputy prime ministers) have had multiple meetings with UN delegations and other international nongovernmental organizations to highlight the need for more assistance and ensure a fair and transparent distribution of humanitarian aid: "We are fully committed to supporting the humanitarian assistance of the UN and other donor organizations. We assure the full security, access, flexible policies, and enabled environment for the donor community in continuing their humanitarian assistance."[38] Almost every tweet posted about the meetings of acting Taliban Foreign Minister Amir Khan Muttaqi with representatives of foreign countries shows that humanitarian assistance has been on the agenda. The Taliban have appointed a ministerial-level committee as a focal point to coordinate humanitarian response with implementing partners, and they have developed an aid distribution procedure.[39] Their posts indicate that

---

[35] Inamullah Samangani [@HabibiSamangani] (Deputy Spokesperson for the IEA), "The aggressors, investing more than $10 billion, failed to eliminate opium, but the Islamic Emirate will do this task just by a decree from Amir ul-Mumineen. Inshallah," Twitter post, April 3, 2022.

[36] Bilal Sarwary [@bsarwary] (freelance journalist), "Sharp increase in Opium prices across Helmand province," Twitter post, May 1, 2022.

[37] VOA [Voice of America] Dari, "Taliban Control Huge Amount of Opium Confiscated by the Previous Regime," video, April 22, 2022.

[38] Arg Presidential Palace [@ARG_1880] (official Twitter account), "Mawlawi Abdul Salam Hanafi: We are fully committed to supporting the humanitarian assistance of the UN and other donor organizations. we assure the full security, access, flexible policies, and enabled environment for donor community in continuing their humanitarian assistance," Twitter post, January 18, 2022.

[39] Inamullah Samangani [@HabibiSamangani] (Deputy Spokesperson for the IEA), "2/2 The acting Minister of Economy thanked UN and the donor countries. And explained the new procedure approved by the IEA cabinet to coordinate humanitarian efforts, ensure transparency in aid distribution, and promised support for the humanitarian work," Twitter post, January 31, 2022.

they have been persistent in asking aid organizations to coordinate their efforts with the relevant authorities of the de facto government.

Unlike the positive atmosphere on Twitter, some Taliban officials have elsewhere questioned the transparency of the aid distribution and criticized aid organizations, including the UN agencies, for not delivering on their promises and for being untransparent.[40] In their messaging to the public, the Taliban portray humanitarian assistance as a kind gesture of aid organizations and donor countries but not one that is critical to addressing economic hardship. Instead, they stress that their economic policies concentrating on agriculture and mining are the proper solutions for the economic crises. The state media are silent about humanitarian efforts led by nongovernmental organizations and broadcast only the humanitarian response of state entities, such as the Afghan Red Cross Society.[41] However, it is worth noting that state news agencies regularly report the arrival of humanitarian money that is being deposited in the corporate accounts of aid agencies in a private commercial bank in Kabul.[42]

Comparing Economic Tweets Across the Languages

In the period covered by this report, the sample Twitter accounts tweeted 74 percent of the economic tweets in the local languages—Pashto and Dari/Farsi—and the rest in English. Economic growth fueled by investment in mining and agriculture sectors, curbing rampant corruption, and improving the security situation are the common themes of the tweets across the languages. Nonetheless, the Taliban seem to tailor economic messaging primarily to the local audience. Accordingly, the communications in the local languages are detailed (highlighting activities related to economics at the policy and operational levels), and they are shared through local media and Twitter. For example, the state news agencies extensively cover the economic news in the local languages. And the Twitter account associated with the office of the First Deputy Prime Minister, which handles economic affairs, posts most of the tweets in Pashto and Dari/Farsi and translates some of them into English.

Overall, variation in the tone and the content of the economic tweets across the languages are less pronounced. The leadership messaging about the economy is usually translated from Pashto into Dari/Farsi and sometimes into English, but it is unclear how tweets are selected for translation and wider dissemination. Tweets about domestic economic policies and activities are primarily in the local languages and generally come from the accounts of the first

---

[40] TOLOnews, March 14, 2022a.

[41] Bakhtar News Agency [@BakhtarNA] (state news agency), "Food items were distributed to more than 7500 families affected by war and drought in the Aqcha district of the Juzjan province. Mawlawi M. Zarif Fayez, the Juzjan Economic Director, said the assistance packages included flour, cooking oil, and peas sponsored by WFP," Twitter post, April 13, 2022.

[42] Bakhtar News Agency [@BakhtarNA] (state news agency), "The Da Afghanistan Bank, central bank, authorities say that as part of the humanitarian aid, 32 million US dollars arrived in Kabul and deposited to the Afghanistan International Bank," Twitter post, March 20, 2022.

Deputy Prime Minister's Office and the deputy spokesperson. The MoFA spokesperson primarily posts English tweets on regional economic cooperation and humanitarian aid.

To sum up, the Taliban use Twitter and other news outlets to highlight their efforts to improve the crumbling economy. Across all the languages we explored, they strive to paint a promising picture of economic growth, improved economic relations with regional countries, and attainment of self-sufficiency, though this picture does not always reflect reality. Table 3.1 lists the top producers of tweets related to the economy.

TABLE 3.1

Ranking of Tweets by Top Producers of Economic Messaging

Name	Title	Language by Frequency from Highest to Lowest
Ahmadullah Muttaqi	Assistant Chief of Staff to the Prime Minister, Deputy Director General Public and Strategic Affairs Office	1. Pashto 2. Dari/Farsi 3. English
Al Emarah English	Official Taliban English Twitter account	1. English
Abdul Wahid Aryan	Director of Bakhtar News Agency	1. English 2. Pashto 3. Dari/Farsi
Bilal Karimi	Deputy Spokesperson for the IEA	1. Pashto 2. Dari/Farsi
Office of Mullah Ab Ghani Barader	First Deputy Prime Minister	1. Dari/Farsi 2. Pashto 3. English
Inamullah Samangani	Deputy Spokesperson for the IEA	1. Dari/Farsi 2. Pashto
Hafiz Zia Ahmad	Deputy Spokesman and Assistant Director of Public Relations, MoFA	1. Pashto 2. Dari/Farsi
Mohammad Naeem Wardak	Spokesman of the Political Office	1. Pashto 2. Dari/Farsi 3. Arabic
Muhammad Jalal	Close to Taliban, but no official title	1. English 2. Pashto 3. Dari/Farsi
MoFA	Official Twitter account	1. Dari/Farsi 2. Pashto 3. English
Abdul Qahar Balkhi	MoFA Spokesperson	1. English 2. Dari/Farsi 3. Pashto
Suhail Shaheen	Taliban Permanent Representative Nominee to UN and Head of Political Office, former Negotiations Team's Member	1. English 2. Pashto 3. Dari/Farsi
Ahmadullah Wasiq	RTA Director	1. Pashto 2. Dari/Farsi
Zabihullah Mujahid	Deputy Minister of Culture and Information and Spokesperson	1. Pashto 2. Dari/Farsi

CHAPTER 4

External Relations and Connections to Regional Militant Groups

After the Taliban regained control of the government on August 15, 2021, a major question was how the new IEA would conduct relations with the world. Grounds for skepticism were not unfounded; during the previous Taliban regime (1996–2001), Afghanistan was virtually isolated, and the first IEA was recognized by only three countries: Pakistan, Saudi Arabia, and the United Arab Emirates. The Taliban at that time also hosted many militant and terrorist groups, most notably al Qaeda, whose presence in Afghanistan and involvement in the September 11, 2001, attacks in the United States provided the casus belli for the War on Terror and the subsequent toppling of the Taliban government in Operation Enduring Freedom.

Although questions remain about the Taliban's commitment to combating terrorism and expelling militants, they appear to be highly interested in maintaining good diplomatic relations across the world and have robustly engaged a multitude of actors—from individual countries to humanitarian organizations to multinational bodies like the European Union. This focus is reflected by the messaging output of the Taliban's media team and spokespersons with tweets discussing external relations in Pashto, Dari/Farsi, Arabic, and English. The Taliban's messaging is uniform and noncontradictory overall, and the media team does not appear to use languages selectively to communicate different messages about external relations. However, the languages used appear to vary somewhat depending on the account and the other parties and/or countries being mentioned. This section identifies some of the common themes that have emerged in the Taliban's externally focused messaging and patterns of interest, which include both interesting messaging and noteworthy omissions.

In this section, externally focused messaging has been divided into two categories: international relations and regional militant groups.

Characterizing Taliban Messaging on International Relations

The majority of Taliban messaging that focuses on international relations is procedural in nature. The Taliban media team extensively covers foreign visits by Taliban officials and meetings with foreign dignitaries at home and abroad. On occasion, some spokespersons also offer comments on events in the world that affect Afghanistan by sharing the IEA's official statement on the matter.[1]

Analysis of Taliban tweets on this topic revealed several country-specific messaging patterns. Some of these patterns, discussed later, may be indicative of deliberate Taliban policy toward those entities and may therefore provide an insight into the IEA's foreign policy.

Some of the countries discussed in this section have long and complex relations with the Taliban and any attempts to discuss these ties in adequate detail using only Twitter would be inadequate. To remain within the scope of this study, this section will focus mainly on information sourced from Taliban Twitter content during the period for which tweets were collected.

Emphasis on Ties with Russia and China

Since August 2021, the Taliban media team has released a considerable volume of content covering meetings between IEA leaders and Chinese officials. These range from high-level meetings between the Chinese Foreign Minister Wang Yi and senior IEA officials (such as Deputy Prime Minister Mullah Baradar and Acting Foreign Minister Amir Khan Muttaqi) to other discussions between Minister for Mining Sheikh Delawar with representatives of MCC, the state-owned Chinese mining company. Common topics of discussion in these meetings are trade, transit, humanitarian aid, and economic development (particularly in the mining sector).[2] In the first few months after the IEA came to power, another common theme was security, with a regular assurance that Afghan soil would not be used against anyone, and a reciprocal assurance from the Chinese side that China espouses the principles of sovereignty and noninterference.[3]

Similarly, the media team has extensively covered meetings of Deputy Prime Minister Mullah Baradar and Acting Foreign Minister Amir Khan Muttaqi with Russian Special Envoy to Afghanistan Zamir Kabulov. In general, these meetings have focused on trade, economic investments, humanitarian aid, and security assistance.[4] In February 2022, however, the IEA

---

[1] In this chapter, we draw distinctions between the *Taliban* (individual of the party in charge who are posting on Twitter) and the *IEA* (the overall Afghanistan government about which they are tweeting).

[2] Abdul Qahar Balkhi [@QaharBalkhi] (MoFA Spokesperson), "Afghan Foreign Minister Mawlawi Amir Khan Muttaqi welcomed Chinese Foreign Minister Wang Yi to Kabul in a special visit to Afghanistan. The Foreign Ministers met in Storai Palace-MoFA to discuss political, economic & transit issues, air corridor, dried fruit export, educational," Twitter post, March 24, 2022.

[3] Inamullah Samangani [@HabibiSamangani] (Deputy Spokesperson for the IEA), "He assured that China would not interfere in Afghanistan's internal affairs. 'Based on the experience of the past 20 years, we call on the international community to allow Afghans to form their own governments and not to try to impose foreign ideas on Afghans,' he said. 2/4," Twitter post, January 15, 2022.

[4] Hafiz Zia Ahmed [@HafizZiaAhmed1] (Deputy Spokesman and Assistant Director of Public Relations, Ministry of Foreign Affairs), "Today, Russian President's Special Envoy Zamir Kabulov led a delegation to Kabul and met with Foreign Minister Maulvi Amir Khan Mottaki. The meeting focused on strengthening political, economic, transit and regional ties. Mr. Kabulov said the policy of the Islamic Emirate is balanced, regional and global," Twitter post, March 24, 2022.

declared neutrality in the Ukraine-Russia conflict and Taliban spokespersons rejected Russian reports of Afghan elements being involved in stoking unrest in Kazakhstan during the mass protests in January this year.[5] This may be an indicator of residual tensions in the relationship.

Heavy Emphasis on Ties with Northern Neighbors but Tajikistan Is a Notable Omission

After the establishment of the IEA government, Taliban leaders began to robustly engage with two of Afghanistan's northern neighbors, Turkmenistan and Uzbekistan. The Taliban media team has diligently documented these diplomatic efforts.

Since August 2021, Deputy Prime Minister Mullah Baradar and Acting Foreign Minister Amir Khan Muttaqi have met several times with Turkmenistan Foreign Minister Raşit Meredow and Deputy Minister Vepa Hajiyev, often with accompanying delegations of officials on both sides. Trade, infrastructure, energy, and regional integration dominated these discussions, with an emphasis on the TAPI pipeline and Turkmenistan Afghanistan Pakistan (TAP) Interconnection Program projects.[6] Other Taliban officials have also met their Turkmen counterparts, often for specific purposes, such as Defense Minister Mullah Yaqoob meeting with Turkmen officials in October to discuss security measures along the common border and special security measures for the TAPI pipeline project.[7] There does appear to be some discord in the messaging surrounding Turkmenistan, however. On January 3, 2022, Deputy Spokesperson Inamullah Samangani rejected reports of conflict on the Afghan-Turkmen

---

[5] Muhammad Jalal [@MJalal313] (no official title), "Statement of MoFA, IEA on Russia-Ukraine conflict: The Islamic Emirate of Afghanistan, in line with its foreign policy of neutrality, calls on both sides of the conflict to resolve the crisis through dialogue and peaceful means. @QaharBalkhi," Twitter post, February 25, 2022.

Also see "Allegations of Afghan Interference in Kazakhstan's Insecurity Are Unfounded," *Voice of Jihad* (Al Emarah English), February 2, 2022).

[6] Zabihullah Mujahid [@Zabehulah_M33] (Deputy Minister of Culture and Information and Spokesperson), "Important: An important meeting was held by the leaders of the Islamic Emirate with a delegation led by the Minister of Foreign Affairs of Turkmenistan Rashid Muradov at the Presidential Palace today. Economic, security, trade and humanitarian assistance were discussed at the meeting," Twitter post, October 30, 2021.

Ministry of Foreign Affairs—Afghanistan [@MoFA_Afg] (official account of the Ministry of Foreign Affairs of Afghanistan), "IEA Deputy Foreign Minister Alhaj Sher Mohammad Abbas Stanekzai met today with Mr. Wafa Khadzhiev, Deputy Foreign Minister of Turkmenistan. Mr. Khadzhiev expressed satisfaction with the overall security situation in Afghanistan, saying Turkmenistan would start work on TAPI," Twitter post, January 8, 2022.

[7] Zabihullah Mujahid [@Zabehulah_M33] (Deputy Minister of Culture and Information and Spokesperson), "Foreign Minister of Turkmenistan Muradov and his accompanying delegation had a special meeting with the Minister of Defense of the Islamic Emirate of Afghanistan Maulvi Mohammad Yaqub Mujahid and other officials at the Presidential Palace. The meeting discussed in detail the security situation on the border between the two countries and in particular the security measures for the TAPI project," Twitter post, October 30, 2021.

border.[8] But during an interview in February, Defense Minister Mullah Yaqoob claimed that the dispute had arisen from the mistreatment of Afghan citizens by Turkmen border security forces, which resulted in the death of one Afghan citizen.[9] Overall however, the IEA appears to be committed to maintaining positive ties with both Uzbekistan and Turkmenistan.

Similarly, Taliban leaders have met numerous times with Uzbek officials, most prominently Foreign Minister Vladimir Norov and Special Representative of the President of Uzbekistan to Afghanistan Ismatulla Irgashev. Common themes in these meetings are transit, trade, humanitarian assistance, energy, and infrastructure, with a greater emphasis on power, given Uzbekistan's status as one of Afghanistan's major electricity suppliers.[10] Regional integration is also mentioned, such as a railway from Termez, Uzbekistan, to Peshawar, Pakistan, through Mazar-e-Sharif and Kabul in Afghanistan.[11] Overall, bilateral ties between Uzbekistan and the IEA appear to be cordial. However, Uzbekistan has yet to return the aircraft and other military equipment taken there by personnel fleeing the republic in August 2021, despite several requests to do so.[12] This could be a source of tension in the relationship.

Tajikistan (Afghanistan's third neighbor in the north), has by contrast been virtually absent from the Taliban's media output. The only major media team statement in connection with Tajikistan was in January, when Taliban Deputy Spokesperson Inamullah Samangani rejected Tajikistan's claims about the presence of terrorist camps along the Afghan-Tajik bor-

---

[8] Inamullah Samangani [@HabibiSamangani] (Deputy Spokesperson for the IEA), "Ministry of Foreign Affairs: Reports of a dispute on the Afghan-Turkmen border are false. There is no problem between us and Turkmenistan's neighbor, and we want to have positive and constructive relations diplomatically with the friendly country of Turkmenistan and other countries of the world," Twitter post, January 4, 2022.

[9] RTA World, "Mawlawi Mohammad Yaqub Mujahid Acting Defense Minister Exclusive Interview with English Subtitles," video, February 15, 2022.

[10] Ahmadullah Wasiq [@WasiqAhmadullah] (RTA Director), "Mawlawi Abdul Kabir, political deputy of the IEA, said today in a meeting with Uzbekistan's special envoy for Afghanistan: 'For recognition, we have met all the conditions of the world and the world must continue to provide humanitarian assistance to Afghans without conditions, all problems should be solved through discussions instead of pressure,'" Twitter post, November 30, 2021.

[11] Abdul Qahar Balkhi [@QaharBalkhi] (MoFA Spokesperson), "Afghan Foreign Minister Mawlawi Amir Khan Muttaqi met this evening with Uzbek delegation led by Minister of Transport Mr. Ilkhom Makhkamov. The meeting focused on constructive discussions on bilateral relation, the Uzbekistan delegation assured Minister Muttaqi of commencing," Twitter post, December 18, 2021.

Abdul Qahar Balkhi [@QaharBalkhi] (MoFA Spokesperson), "Uzbekistan-Afghanistan-Pakistan railway in the coming spring, saying Afghans will be trained by Uzbekistan on railway operation," Twitter post, December 18, 2021.

[12] Office of First Deputy Prime Minister [@FDPM_AFG] (official Twitter account), "1/6: The First Deputy Prime Minister met with Esmatullah Agha, Special Representative of the President of Uzbekistan for Afghanistan! During the meeting, which took place between the two sides in the Citadel today, Tuesday, political and economic issues were discussed," Twitter post, November 30, 2021.

Office of First Deputy Prime Minister [@FDPM_AFG] (official Twitter account), "5/6: He added that he would decide on helicopters and all military equipment transferred to their country in accordance with international law," Twitter post, November 30, 2021.

der.[13] Tajikistan has been much cooler toward the IEA than the other Central Asian republics and it reportedly hosts and supports elements of the anti-Taliban armed opposition.[14] The absence of Tajikistan in the Taliban's media content is likely a reflection of mutual animosity.

Emphasis on Ties with Qatar, Iran, and Turkey; Little Attention Toward Saudi Arabia and the United Arab Emirates

Saudi Arabia and the United Arab Emirates were two of only three countries to recognize the first IEA; Turkey supported General Abdul Rashid Dostum, one of the most prominent anti-Taliban leaders, and Iran supported the anti-Taliban Northern Alliance led by Ahmad Shah Massoud. This time around, the Taliban media team's coverage appears to indicate that the IEA has minimal contacts with its former Gulf allies. Although the IEA has condemned Houthi attacks on Saudi Arabia and the United Arab Emirates,[15] its meetings with delegations from those countries have mostly been focused on humanitarian aid and economic investment, eschewing potentially divisive discussion on such topics as cultural exchanges, requests for facilitation of students and businessmen, and other issues discussed frequently with Turkey and Qatar.[16]

In contrast, Taliban messaging indicates efforts to build strong relationships with Turkey, Iran, and Qatar. In Turkey's case, private individuals are playing a leading role in building the relationship and Taliban officials have met several delegations of Turkish businessmen, investors, and humanitarian aid workers to discuss investment and humanitarian aid.[17] Meetings

---

[13] Inamullah Samangani [@HabibiSamangani], Deputy Spokesperson for the IEA, "The claim of the President of Tajikistan that many camps have been set up in Afghanistan in the border areas with Tajikistan for destructive activities is not true and we reject it seriously. The Islamic Emirate assures all neighboring countries that our borders are safe and . . . 1/2," Twitter post, January 11, 2022.

Inamullah Samangani [@HabibiSamangani], Deputy Spokesperson for the IEA, "There is no threat to any country from our territory, including Tajikistan. Some circles and fugitives and biased people transmit false information to neighboring countries and the world, which is never true. 2/2," Twitter post, January 11, 2022.

[14] Vinay Kaura, "Tajikistan's Evolving Relations With the Taliban 2.0," Middle East Institute, December 1, 2021.

[15] Abdul Qahar Balkhi [@QaharBalkhi] (MoFA Spokesperson), "The Ministry of Foreign Affairs of IEA is saddened to learn about the rocket attacks on oil facilities and civilian targets in Jeddah-KSA, and condemns attacks on civilian targets. IEA believes that such acts threaten regional peace and stability and should be ceased," Twitter post, March 26, 2022.

[16] Abdul Qahar Balkhi [@QaharBalkhi] (MoFA Spokesperson), "Following his arrival in Kabul today, UAE Minister of Federal Authority for Identity, Citizenship, Customs & Ports Security Mr. Ali Mohammed bin Hammad Al Shamsi discussed issues of importance with IEA Foreign Minister Mawlawi Amir Khan Muttaqi," Twitter post, December 28, 2021.

[17] Abdul Qahar Balkhi [@QaharBalkhi] (MoFA Spokesperson), "On Friday a senior IEA delegation led by Foreign Minister Mawlawi Amir Khan Muttaqi met Yunus Sezer, head of Turkey's Disaster & Emergency Management Presidency & accompanying delegation. Mr. Sezer said they were ready to help Afghanistan in training & humanitarian assistance," Twitter post, October 17, 2021.

with Turkish government officials have tended to focus on trade, transit for Afghan citizens, and educational opportunities for Afghan students in Turkish institutions.[18]

Qatar has hosted the Taliban's political office since 2013, and Doha is where the Taliban and the United States negotiated the terms of the U.S. withdrawal, culminating in the Doha Accords. Since the Taliban took power, Qatar has continued to serve as a bridge between the IEA and the international community, and the IEA has had several postwithdrawal meetings with representatives of the United States, European countries, and certain other nations, such as Japan. The Taliban also appear to view bilateral ties with Qatar positively and the two sides have met regularly to discuss humanitarian aid, economic development, and the education sector.[19]

The IEA has also been in regular contact with Iran; Foreign Minister Muttaqi and Deputy Prime Minister Baradar have met several times with senior Iranian officials, including Iranian Foreign Minister Hossein Amir-Abdollahian and Special Representative of Iran for Afghanistan Hassan Kazemi Ghomi. Aside from the common themes of trade, improvement in bilateral ties, and economic cooperation,[20] the topic of Afghan refugees and transit facilities for them is also often discussed.[21] However, there are tensions in the relationship: There have been at least two incidents of border clashes between the countries, though Taliban media have tried to downplay these incidents by characterizing them as misunderstandings.[22]

---

[18] Abdul Qahar Balkhi [@QaharBalkhi] (MoFA Spokesperson), "The Turkish Foreign Minister said Turkish companies and traders would visit Afghanistan. Minister Muttaqi said security and facilities would be provided for Turkish traders and humanitarian organization and all impediments would be removed," Twitter post, December 19, 2021.

[19] Ministry of Foreign Affairs—Afghanistan [@MoFA_Afg], Official account of the Ministry of Foreign Affairs of Afghanistan, "A meeting between the Islamic Emirate delegation led by Acting Foreign Minister Mawlawi Amir Khan Muttaqi was held with Qatari officials from several civil service and education institutions. The meeting focused on humanitarian situation in Afghanistan, dynamics of higher," Twitter post, February 15, 2022.

Ministry of Foreign Affairs—Afghanistan [@MoFA_Afg], Official account of the Ministry of Foreign Affairs of Afghanistan, "education, capacity development, humanitarian aid, and providing scholarships to Afghan students. The Qatari officials pledged to assist in various areas, provide training opportunities for Afghan students, and deliver aid to Afghanistan through Qatari Red Crescent and," Twitter post, February 15, 2022.

[20] Abdul Qahar Balkhi [@QaharBalkhi], MoFA Spokesperson, "IEA FM Mawlawi Amir Khan Muttaqi met Iranian FM Mr. Hossein Amir Abdollahian during the extraordinary OIC session on Afghanistan. The two sides talked on current security and humanitarian situation as well as trade, economy, and political relations between the two countries," Twitter Post, December 19, 2021.

[21] Ahmadullah Wasiq [@WasiqAhmadullah] (RTA Director), "His Excellency, Mr. Muttaqi expressed the readiness of the Islamic Emirate in these areas and also demanded that special attention should be paid to the solution of the refugee problem in Iran and steps should be taken to facilitate the travelers to and from the border," Twitter post, October 23, 2021.

[22] Zabihullah Mujahid [@Zabehulah_M33] (Deputy Minister of Culture and Information and Spokesperson), "The incident between the Afghan and Iranian border guards in Nimroz was brought under control. The incident took place between the Afghan and Iranian border guards in Nimroz province," Twitter post, December 1, 2021.

Mixed Relationship with Pakistan, Minimal Contact with India

Pakistan was the third country, along with Saudi Arabia and the United Arab Emirates, to recognize the first IEA as a sovereign government. This time around, Pakistan has refrained from conferring recognition on the IEA despite the passage of several months since it came to power. However, in general, relations between the countries appear to be cordial, with IEA officials having met several times with Foreign Minister of Pakistan Bilawal Bhutto Zardari, Pakistani Ambassador in Kabul Obaid Ur Rehman Nizamani, and even Prime Minister Imran Khan while he was in office. During these meetings, the main topics of discussion have been trade and transit issues, particularly for Afghan citizens and traders, and of the two major border crossings at Chaman-Spin Boldak and Torkham, at least one of which is referenced in most meetings in the context of transit issues and visas.[23] Taliban spokespersons have also appreciated Pakistani assistance measures, such as removal of sales tax on Afghan fruit exports and the provision of humanitarian aid,[24] and have condemned terrorist attacks in Pakistan.[25]

However, border tensions remain. There were at least two distinct incidents along the border in January and February, and in April, Pakistan launched air strikes within Afghan territory.[26] Pakistan claimed to have struck terrorist hideouts responsible for attacks on its

---

[23] Inamullah Samangani [@HabibiSamangani] (Deputy Spokesperson for the IEA), "Foreign Minister Maulvi Amir Khan Muttaqi, Trade and Finance Ministers, and their accompanying delegation met with Pakistani Prime Minister Imran Khan, cabinet members and other high-ranking officials. The Foreign Minister called for facilities to be provided at Torkham and Chaman Gates to facilitate the transit of Afghan goods 1/2," Twitter post, November 12, 2021.

Zabihullah Mujahid [@Zabehulah_M33] (Deputy Minister of Culture and Information and Spokesperson), "Qureshi: Immediate visas will be issued to Afghan patients in Torkham and will facilitate their travel," Twitter post, October 21, 2021.

[24] Zabihullah Mujahid [@Zabehulah_M33] (Deputy Minister of Culture and Information and Spokesperson), "1/3. Press Release of the Ministry of Foreign Affairs: As a result of the efforts, negotiations and communication of the Ministry of Foreign Affairs, the Islamic Emirate of Pakistan, in order to solve issues regarding the export of fresh fruit," Twitter post, September 25, 2021.

Zabihullah Mujahid [@Zabehulah_M33] (Deputy Minister of Culture and Information and Spokesperson), "has lifted 17% tax (Sale tax) that had been imposed some time back with effect from September 24, 2021. The Ministry of Foreign Affairs of the Islamic Emirate of Pakistan welcomes this step of the Islamic Republic of Pakistan," Twitter post, September 25, 2021.

[25] Zabihullah Mujahid [@Zabehulah_M33] (Deputy Minister of Culture and Information and Spokesperson), "As part of its assistance to Afghans, Pakistan has pledged 50,000 tonnes of wheat to Afghanistan, with the first batch of 3600 tonnes arrived in Nangarhar yesterday. We welcome this assistance from our brotherly country Pakistan and ask for more cooperation," Twitter post, March 1, 2022.

Zabihullah Mujahid [@Zabehulah_M33] (Deputy Minister of Culture and Information and Spokesperson), "Sympathy: We condemn the bombing of a mosque in Peshawar, Pakistan. There is no justification for attacking civilians and worshipers. We express our deepest condolences to all the victims of the incident," Twitter post, March 4, 2022.

[26] Abdul Qahar Balkhi [@QaharBalkhi] (MoFA Spokesperson), "Recently a few of incidents have taken place along Durand line between Afghanistan and Pakistan that have given rise to the need for authorities

soldiers along the border, but Taliban spokespersons denied this and condemned the strikes, calling on issues to be solved by dialogue instead.[27]

Since the IEA's establishment, India has not been featured much in the Taliban's Twitter content, and most statements within this limited sample have been expressions of gratitude for humanitarian assistance.[28] However, the trajectory of this relationship appears to be changing with the reopening of New Delhi's embassy in Kabul.[29]

Positive Statements Toward Engagement with European Nations and Japan

With the end of the war, the IEA seem to be interested in positive engagement with several former enemies. Since August, IEA leaders have met with ambassadors and special representatives of Japan and the European Union, as well as with officials from individual members of the European Union. Among these countries, Germany, the United Kingdom, Japan, the Netherlands, and Norway stand out as having the greatest amount of coverage in the Taliban media's output. Common topics of discussion include humanitarian aid, improvement in bilateral ties, and economic cooperation and investment.[30] IEA officials have also used some

---

of the two sides to address the problem," Twitter post, January 4, 2022.

Zabihullah Mujahid [@Zabehulah_M33] (Deputy Minister of Culture and Information and Spokesperson), "A local incident took place in Spin Boldak area of Kandahar with Pakistani guards. Unfortunately, the first shots were fired by Pakistani guards. Leaders on both sides have been informed and the situation is now under control. We will do a thorough investigation into why this happened," Twitter post, February 24, 2022.

[27] Zabihullah Mujahid [@Zabehulah_M33] (Deputy Minister of Culture and Information and Spokesperson), "1/2- The Islamic Emirate of Afghanistan strongly condemns Pakistan's attacks on refugees in Khost and Kunar. IEA calls on the Pakistani side not to test the patience of Afghans on such issues and not repeat the same mistake again otherwise it will have bad consequences," Twitter post, April 16, 2022.

[28] Inamullah Samangani [@HabibiSamangani] (Deputy Spokesperson for the IEA), "As a result of India's humanitarian assistance to the Afghans, six tons of various medicines arrived in Kabul from that country. India has already sent 500,000 doses of corona vaccine to Afghanistan. The Islamic Emirate of Afghanistan would like to thank India for its humanitarian assistance," Twitter post, January 7, 2022.

[29] As of June 2022, India had reopened its embassy in Kabul and there was a high-level meeting between senior IEA leadership and Indian officials. IEA officials have also invited India to complete its development projects in Afghanistan and looked favorably on strengthened relations between the two sides. However, because these developments occurred outside the timeline of this study, they are mentioned as an update here rather than in the main body.

[30] Abdul Qahar Balkhi [@QaharBalkhi] (MoFA Spokesperson), "IEA Foreign Minister H.E. Mawlawi Amir Khan Muttaqi and accompanying delegation met with the UK Charge d'Affaires for Afghanistan Mission Mr. Martin Longden & discussed the humanitarian, economic & security situation of Afghanistan," Twitter post, November 28, 2021.

Abdul Qahar Balkhi [@QaharBalkhi] (MoFA Spokesperson), "IEA delegation assured them about security and emphasized that Afghanistan seeks positive relations with the world, similarly urging transparent and urgent health assistance to Afghanistan. In the end, both sides agreed to continue such dialogue," Twitter post, November 28, 2021.

of these meetings to urge Western countries to persuade the United States to lift sanctions and unfreeze Afghanistan's foreign reserves.[31]

Mixed Attitudes About the United States

Taliban messaging exhibits mixed sentiments regarding the United States. In general, messaging regarding the United States is negative, with several spokespersons condemning U.S. actions during the war or afterward. Some major points of criticism include the freezing of the Afghan national reserves, President Joe Biden's statement implying that Afghanistan was not united, and the administration's decision in January 2022 to fence off $3.5 billion of the Afghan foreign reserves as potential compensation for victims of the September 11, 2001, terror attacks, subject to the outcome of legal cases, a move denounced by IEA representatives as "confiscation."[32] In every meeting with U.S. officials, Taliban leaders make certain to raise the issue of frozen reserves and call for them to be unfrozen and handed over to the IEA.[33]

On the other hand, the media team has also covered several meetings between senior Taliban officials and the U.S. Special Representative to Afghanistan Thomas West that ended on a positive note, and the IEA has affirmed many times that it desires to have good relations with all

---

[31] Suhail Shaheen [@suhailshaheen1] (Taliban Permanent Representative Nominee to UN and Head of Political Office, former Negotiations Team's Member), "1/2 I met His Excellency Markus Putzel, German ambassador to Afghanistan in Doha today and discussed with him a number of issues including removal of current sanctions, needs for reconstruction and development projects in the country," Twitter post, March 22, 2022.

[32] Suhail Shaheen [@suhailshaheen1] (Taliban Permanent Representative Nominee to UN and Head of Political Office, former Negotiations Team's Member), "1/2 Reserve of Da Afghanistan Bank does not belong to governments or factions but it is property of the people of Afghanistan. It is only used for implementation of monetary policy, facilitation of trade and boosting of financial system of the country. It is never intended to be," Twitter post, February 12, 2022.

Suhail Shaheen [@suhailshaheen1] (Taliban Permanent Representative Nominee to UN and Head of Political Office, former Negotiations Team's Member), "2/2 used for any other purpose rather than that. It's freezing or disbursement unilaterally for any other purpose is injustice and not acceptable to the people of Afghanistan," Twitter post, February 12, 2022.

Abdul Qahar Balkhi [@QaharBalkhi] (MoFA Spokesperson), "The Ministry of Foreign Affairs of IEA strongly rejects remarks by President @POTUS asserting Afghanistan is incapable of unity," Twitter post, January 20, 2022.

Muhammad Jalal [@MJalal313] (no official title), "American governments terrorized Afghans for 20 years and now your boss @POTUS is stealing the money belonging to innocent Afghans," Twitter post, February 11, 2022.

[33] Hafiz Zia Ahmed [@HafizZiaAhmed1] (Deputy Spokesman and Assistant Director of Public Relations, Ministry of Foreign Affairs), "The two sides held detailed discussions on current political & economic situation in Afghanistan. Both sides agreed that the Afghan Central Bank's USD $3.5 billion unfrozen assets from the US bank shall in no circumstances be given to charity organizations. The Afghan side," Twitter post, March 11, 2022.

countries—including, specifically, the United States.[34] Taliban spokespersons have also praised what they consider to be positive steps from the United States, including the Treasury Department's decision in December 2021 to facilitate the transfer of humanitarian aid to Afghanistan, and calls by certain members of Congress for positive engagement with Afghanistan.[35]

Emphasis on Economic and Trade Issues, Energy, and Infrastructure, Good Ties with All

Trade, economic cooperation and development, and humanitarian aid are discussed most frequently in meetings between IEA leaders and foreign officials. These topics come up in virtually every meeting, regardless of the country. Some issues are related to a specific country and are only raised with those parties: For instance, transit issues related to the crossings at Torkham and Chaman are frequently discussed in meetings with Pakistani officials, electricity and railways are common themes in meetings with Uzbek officials, and the TAP Interconnection Program is often mentioned in discussions with Turkmen dignitaries.

A common refrain of IEA officials in meetings is that they seek positive engagement with the world,[36] economic cooperation to make Afghanistan a hub of regional connectivity,[37] and good relations with all countries and neighbors, including past enemies, such as the United States. However, these statements also lay down an expectation of reciprocity, asking for noninterference in Afghanistan's affairs.[38]

Engagement with Humanitarian Organizations

Several different international organizations and nonprofits are working to address the adverse humanitarian situation in Afghanistan. The Taliban have taken care to engage with

---

[34] Abdul Qahar Balkhi [@QaharBalkhi], MoFA Spokesperson, "Two-day dialogue between delegations of the Islamic Emirate and USA in Doha," Twitter post, October 10, 2021.

[35] Abdul Qahar Balkhi [@QaharBalkhi] (MoFA Spokesperson), "Remarks by MoFA spokesperson regarding recent US action: MoFA of the IEA welcomes recent decision by US Treasury Department allowing US gov agencies along with international and nongovernmental organizations & banks to facilitate flow of food & medicine to the IEA," Twitter post, September 25, 2021.

Abdul Qahar Balkhi [@QaharBalkhi] (MoFA Spokesperson), "We appreciate step by 48 US Congress members who, following our open letter to the Congress, have recommended President Joe Biden (@POTUS) lift sanctions imposed on Afghanistan & unfreeze its Central Bank reserves," Twitter post, December 21, 2021.

[36] Abdul Qahar Balkhi [@QaharBalkhi] (MoFA Spokesperson), "Transcript of Speech by the Deputy Prime Minister, His Excellency Mullah Abdul Ghani Baradar," Twitter post, September 30, 2021.

[37] Abdul Qahar Balkhi [@QaharBalkhi] (MoFA Spokesperson), "Minister Muttaqi said the new government is focused on regional security & connectivity, & wants Afghanistan, as the heart of Asia, to play an important role in strengthening transit, trade, industry & economy among the countries of the region," Twitter post, March 24, 2022.

[38] Abdul Qahar Balkhi [@QaharBalkhi] (MoFA Spokesperson), "Transcript of Speech by the Acting Foreign Minister, H.E. Mawlawi Amir Khan Muttaqi," Twitter post, September 30, 2021.

these groups, and the Taliban media team has extensively covered meetings with representatives and delegations of humanitarian organizations. Common themes in these meetings are expressions of gratitude for the organization's work, requests to assist those in remote areas or those affected by the war,[39] assurances of security and protection,[40] and affirmations of the IEA's readiness to assist in the transparent distribution of aid.[41] In some of these meetings, IEA officials have pushed specialized priorities, such as asking for assistance with tackling the narcotics trade by offering alternatives to incentivize poppy farmers to stop production.[42]

Characterizing Taliban Messaging on Regional Militant Groups

Prior to the fall of the republic, the Taliban were known to have links with several militant groups that were classified as terrorist groups by the United States, Russia, China and/or neighboring countries. These include al Qaeda, the Tehrik-e-Taliban Pakistan (TTP), Islamic Movement of Uzbekistan (IMU), and Turkistan Islamic Party (TIP) among others. The regional affiliate of the Islamic State, the ISKP, also had footholds in the country—but unlike the others, it was and remains hostile to the Taliban.[43]

According to the terms of the Doha Agreement, the peace deal negotiated by the United States and the Taliban in 2020, the latter were required to prevent any groups from using Afghan soil against the United States and its allies, a commitment that the Taliban affirmed by stating that no one would be allowed to use Afghan soil against any country.[44]

---

[39] Inamullah Samangani [@HabibiSamangani] (Deputy Spokesperson for the IEA), "Maulvi Abdul Salam Hanafi expressed his gratitude for the humanitarian aid provided by the United Nations and called for more attention to be paid to the people of remote districts and areas according to their needs. 5/7," Twitter post, February 8, 2022.

[40] Zabihullah Mujahid [@Zabehulah_M33] (Deputy Minister of Culture and Information and Spokesperson), "4/6 and Afghanistan's economy improves somewhat. Maulvi Abdul Salam Hanafi thanked the WFP for its cooperation with Afghanistan in various fields and said, 'We are ready for any kind of cooperation with all the UN agencies and ensure their full security,'" Twitter post, November 7, 2021.

[41] Abdul Qahar Balkhi [@QaharBalkhi] (MoFA Spokesperson), "her voice for the resumption of banking system of Afghanistan. Expressing gratitude to their efforts, Minister Muttaqi assured that the Islamic Emirate was fully prepared to deliver humanitarian aid," Twitter post, October 23, 2021.

[42] Inamullah Samangani [@HabibiSamangani], Deputy Spokesperson for the IEA, "Maulvi Abdul Salam Hanafi called interaction, coordination and cooperation important for development and said: 'The cooperation of the international community in preventing the cultivation, use and trafficking of narcotics and providing alternatives to farmers is important.' He emphasized the need for transparency in the distribution of humanitarian aid 5/7," Twitter post, March 29, 2022.

[43] Seldin, Jeff, "How Afghanistan's Militant Groups Are Evolving Under Taliban Rule," VOA News, March 20, 2022.

[44] Muhammad Jalal [@MJalal313] (no official title), "H.E Khalifa Sirajuddin Haqqani: We will not allow anyone to use Afghan soil against others," Twitter post, March 6, 2022.

Taliban messaging on this issue has been limited—understandably so, considering that the IEA has been eager to downplay any terrorist threats emanating from Afghanistan since its establishment. However, analysis of this content still yields certain observations that may shed light on the IEA's approach to dealing with terrorism.

The Taliban are known to have continued relationships with several militant groups, including al Qaeda, the TTP, IMU, and TIP. Since August 2021, no Taliban spokesperson has mentioned the presence of any of these groups by name. During an interview in February 2022, Defense Minister Mullah Yaqoob rejected claims by Pakistan that there were terrorist hideouts in Afghanistan and that recent attacks on Pakistani troops along the border had been launched from there.[45] The only reference to this subject was on February 7, when MoFA and some spokespersons rejected a report by the UN Security Council Monitoring Group stating that the activity of foreign militant groups was on the rise in Afghanistan.[46]

Consistent Messaging on No Threats Being Allowed to Emanate from Afghanistan

The Taliban media team have amplified several statements from leaders affirming that the IEA was committed to the Doha agreement, that Afghanistan would not pose a threat to anyone, and that the IEA would not allow anyone to use Afghan soil against any country.[47] This messaging has been consistent and is frequently used in meetings with foreign dignitaries.

The Taliban's reliability on this matter has been badly undermined, however, by the assassination of Ayman al-Zawahiri, leader of al Qaeda, in Kabul on July 31, 2022. According to U.S. officials, al-Zawahiri moved to Kabul months after the Taliban's takeover and senior Taliban officials were aware of his presence; even the house he was staying in was owned by an aide of Sarajuddin Haqqani. In the aftermath of the strike, both the Taliban and the United States have traded criticisms about violating the Doha Agreement.[48]

---

[45] RTA World, 2022.

[46] Inamullah Samangani [@HabibiSamangani] (Deputy Spokesperson for the IEA), "Ministry of Foreign Affairs: The report of the UN Security Council Monitoring Group on the increase of foreign groups in Afghanistan is not true," Twitter post, February 6, 2022.

[47] Muhammad Jalal [@MJalal313] (no official title), "H.E Khalifa Sirajuddin Haqqani: We are committed in all our commitments. Afghanistan will no longer be a threat to any of its neighbors and other countries," Twitter post, March 24, 2022.

Abdul Qahar Balkhi [@QaharBalkhi] (MoFA Spokesperson), "said it was not true, adding Afghans consider such rhetoric concerning, provocative, & direct incitement & campaign for subversive groups. Minister Muttaqi said the Afghan government will allow none to use the Afghan soil against another," Twitter post, February 22, 2022.

[48] Matthew Lee, Nomaan Merchant, and Mike Balsamo, "CIA Drone Strike Kills al-Qaida Leader Ayman al-Zawahri in Afghanistan," *PBS News Hour*, August 1, 2022.

Conflicting Messaging on ISKP

ISKP, the Islamic State's regional branch, has posed the toughest security challenge to the Taliban since the establishment of the IEA in August 2021. Taliban messaging about ISKP is limited but mixed.

In the initial months after the change in government, ISKP launched several attacks on Taliban soldiers and civilians. During this period, some Taliban individuals, such as Deputy Minister for Culture and Arts Atiqullah Azizi, asserted that ISKP was being supported by elements of the former government and that the United States and Europe were turning a blind eye.[49]

In January 2022, Aslam Farooqui, a major commander of ISKP, was killed.[50] This coincided with major Taliban operations against the group; from that month onward, the Taliban began to discount ISKP as a threat. During an interview in February 2022, Defense Minister Mullah Yaqoob claimed that ISKP was no longer a threat, a sentiment echoed by the General Directorate of Intelligence in March and by Al Emarah.[51]

Key Sources of External Relations Messaging

The Taliban officials and accounts that RAND identified for this study did not all discuss external relations with the same frequency. Rather, a handful of accounts produced most of the tweets for this topic. Table 4.1 lists these top accounts.

Messaging by Language

In the context of external relations messaging, the Taliban media team and various spokespersons do not appear to use languages selectively to propagate specific messaging. Different spokespersons seem to preferentially use different languages, but several spokespersons (including Qahar Balkhi, Inamullah Samangani, and Zabihullah Mujahid) publish most messaging in multiple languages, and the content across all spokespersons and languages is uniform and noncontradictory.

Pashto is the most common language used by the Taliban media team for external relations messaging. Most spokespersons, except for Inamullah Samangani, use Pashto for all kinds of statements, and for all countries and organizations.

---

[49] Atiqullah Azizi [@at_azizi1] (Deputy Minister for Culture and Arts), "In Afghanistan, wherever ISIS attacks, corrupt officials of the previous regime strongly support them. Bloody, America and Europe have turned a blind eye to it. Such support is neither terrorism nor fundamentalism? !!!," Twitter post, October 23, 2021.

[50] Muhammad Jalal [@MJalal313] (no official title), "Update: Former head of Daesh- ISKP terrorist group, Aslam Farooqi has been killed in Northern Afghanistan. He was at large and involved in crimes against Afghanistan. #Peace," Twitter post, January 16, 2022.

[51] Muhammad Jalal [@MJalal313] (no official title), "GDI: We assure our nation, our neighbors and the world that Daesh is no longer a matter of concern in Afghanistan," Twitter post, March 20, 2022.

TABLE 4.1

Top Producers of External Relations Messaging

Name	Title	Language by Frequency from Highest to Lowest
Abdul Qahar Balkhi	MoFA Spokesperson	1. Pashto 2. English 3. Dari/Farsi
Mohammed Naeem Wardak	Spokesman of the Political Office	1. Pashto 2. Dari/Farsi 3. Arabic
Inamullah Samangani	Deputy Spokesperson of the Islamic Emirate	1. Dari/Farsi 2. English
Zabihullah Mujahid	Deputy Minister of Culture and Information and Spokesperson	1. Pashto 2. Dari/Farsi 3. English
Al Emarah English	English version of the official Taliban publication	1. English

Dari/Farsi is generally the second most common language for IEA media content, though Inamullah Samangani publishes statements almost exclusively in Dari/Farsi with little Pashto content. Most content featuring Iran has a Dari/Farsi version, but there are cases where Iran-focused material has been published only in Pashto and/or English. On the other hand, most messaging related to Pakistan, for instance, is not published in Dari/Farsi, though some content about Pakistan also has a Dari/Farsi version. This makes it difficult to establish a pattern for the use of Dari/Farsi.

Arabic is little used by most of the media team. Mohammad Naeem Wardak is the most prolific publisher in Arabic and even in his case, Arabic tweets account for just around 10 percent of the external relations messaging he has published. Some Taliban spokespersons tweeted Arabic versions of their posts condemning attacks in Saudi Arabia and the United Arab Emirates; aside from that instance, there is no discernible pattern to the use of Arabic. Even in Wardak's case, some of the Arabic tweets pertain to non-Arabic speaking countries, so it is difficult to identify a pattern of usage.[52]

English is commonly used by the major spokespersons (Inamullah Samangani, Qahar Balkhi, and Zabihullah Mujahid), by Deputy Spokesman and Assistant Director of Public Relations for MoFA Hafiz Zia Ahmad, and by one of the major Taliban activists, Muhammad Jalal. Aside from spokespersons, accounts representing major leaders—such as Foreign Minister Muttaqi, First Deputy Prime Minister Mullah Baradar, and Head of the Political Office Suhail Shaheen—also frequently publish in English. However, the Spokesman of the Political Office in Doha, Mohammad Naeem Wardak, rarely uses English.

---

[52] Urdu was also infrequently used by Taliban, with few tweets to review, so we do not include analysis here.

CHAPTER 5

Women's Rights and Education

The new IEA's impact on Afghan women's rights has alarmed both Afghans and members of the international community. Under the previous Taliban regime, and in territories controlled by the Taliban before their recent takeover, women experienced violence, harassment, deprivation of bodily rights, exclusion from the workforce, and restricted access to education and health care.[1] With the Taliban once again in power, a return to this situation, and a loss of gains made for women's rights in the past two decades, is a concern.

In the context of this domestic and international concern, messaging about women's rights has been prominent in the Taliban media team's output. Their tweets in Pashto, Dari/Farsi, Arabic, and English all discuss women's issues with some frequency. However, the mix of discussed themes surrounding women's issues varies across language groups. This section identifies these themes, the figures who discuss them most frequently, and how they differ across languages.

Characterizing Taliban Messaging on Women's Issues

We categorized tweets on women's rights into seven themes. Here, we present descriptions and examples of each.

The IEA Supports Women's Education and Workforce Participation

The Taliban media team often highlights activities or statements by the administration that support women's access to education and workforce participation. For example, the media team heavily promoted the Taliban's intention to reopen schools for girls on March 23, 2022. The following tweets are characteristic of this theme:

> Information from the Ministry of Public Health of the Islamic Emirate of Afghanistan: The Ministry of Public Health of the Islamic Emirate informs all female employees to

---

[1] Cheryl Benard, Seth G. Jones, Olga Oliker, Cathryn Quantic Thurston, Brooke Stearns Lawson, and Kristen Cordell, *Women and Nation-Building*, RAND Corporation, MG-579-IMEY/CMEPP, 2008, pp. 18–22.

attend their duties regularly in the center and provinces. There is no impediment from the Islamic Emirate to carrying out their work.[2]

The decision to open all schools in Afghanistan for all students is because the Islamic Emirate of Afghanistan wants education for all Afghans. Providing education to all Afghans is one of the important goals of IEA.[3]

Five female police officers were employed at the Passport Directorate of Maidan Wardak. The police chief of Maidan Wardak told to Bakhtar News Agency that the female police would work to facilitate the biometric process of female's passports.[4]

The IEA Protects Women's Rights

Taliban messaging emphasizes the regime's desire to ensure that all Afghan women have the rights that Islam grants them. An instance that received significant exposure was the Supreme Leader's special decree on women's rights that included expanded rights for widows, protections for wives in polygynous marriages, and a call for religious scholars to devote attention to women's issues.[5] The following tweets are characteristic of this theme:

Islam has granted a woman the best rights of life and they must be observed and given to all the women.[6]

---

[2] Qari Yousaf Ahmadi [@QyAhmadi21] (Spokesman of the Islamic Emirate of Afghanistan), "Information from the Ministry of Public Health of the Islamic Emirate of Afghanistan: The Ministry of Public Health of the Islamic Emirate informs all female employees to attend their duties regularly in the center and provinces. There is no impediment from the Islamic Emirate to carrying out their work," Twitter post, August 27, 2021.

[3] Muhammad Jalal [@MJalal313] (no official title), "The decision to open all schools in Afghanistan for all students is because the Islamic Emirate of Afghanistan wants education for all Afghans. Providing education to all Afghans is one of the important goals of IEA," Twitter post, March 3, 2022.

[4] Mohammad Naeem Wardak [@IeaOffice] (Spokesman of the Political Office), "Five female police officers were employed at the Passport Directorate of Maidan Wardak. The police chief of Maidan Wardak told to Bakhtar News Agency that the female police would work to facilitate the biometric process of female's passports," Twitter post, March 7, 2022.

[5] Mohammad Naeem Wardak [@IeaOffice] (Spokesman of the Political Office), "1 / 8- Amir al-Mo'menin issued a special decree on women's rights in the name of God: The Supreme Leader of the Islamic Emirate directs all officials of the Islamic Emirate, religious scholars and tribal elders to take serious measures to ensure the following rights of women: 1- The consent of adult girls is necessary during marriage," Twitter post, December 3, 2021.

[6] Mohammad Naeem Wardak [@IeaOffice] (Spokesman of the Political Office), "Islam has granted a woman the best rights of life and they must be observed and given to all the women," Twitter post, March 8, 2022.

"Amr al-Ma'ruf does not mean that a veiled woman, whose whole body is covered except her face, should be covered under the name of Islam and half of her body should be exposed in the beating." Sheikh Khalid Hanafi (Acting Minister)[7]

IEA is fully committed to upholding of all the Sharia rights of the Afghan women. International Women's Day is an opportunity for our Afghan women to demand their legitimate rights. We will protect and defend the rights of our Afghan women, Insha Allah.[8]

Afghan Women Are Happy with the New Regime

Perhaps as a counter to reports of women's rights abuses in Afghanistan, Taliban tweets often share quotations or events that demonstrate women's satisfaction with the IEA regime. The opinions that these women share include support for traditional Islamic policies, such as hijab mandates; rejection of Western portrayals of their situation; and satisfaction with the economic and educational opportunities the Taliban have brought. The following tweets are characteristic of this theme:

A number of women marched in Kabul today with the slogan (Hijab of our honor, Hijab of our adornment, Hijab of our pride). At the end of the march, they declared their full satisfaction and support for the Islamic Emirate.[9]

Afghan women are rejecting those who are trying to represent them in Europe and elsewhere.[10]

Faryab Province: A peaceful women's march: We do not want democracy. We want our rights that Islam has given us. A group of immodest women do not represent Afghan women. Women's right to be educated. And to know. And to operate in the light of

---

7 Muhammad Jalal [@MJalal313] (no official title), "'Amr al-Ma'ruf does not mean that a veiled woman, whose whole body is covered except her face, should be covered under the name of Islam and half of her body should be exposed in the beating.' Sheikh Khalid Hanafi (Acting Minister)," Twitter post, November 17, 2021.

8 Zabihullah Mujahid [@Zabehulah_M33] (Central Spokesperson and Deputy Minister of Culture and Information), "IEA is fully committed to upholding of all the Sharia rights of the Afghan women. International Women's Day is an opportunity for our Afghan women to demand their legitimate rights. We will protect and defend the rights of our Afghan women, Insha Allah," Twitter post, March 8, 2022.

9 Mohammad Naeem Wardak [@IeaOffice] (Spokesman of the Political Office), "A number of women marched in Kabul today with the slogan (Hijab of our honor, Hijab of our adornment, Hijab of our pride). At the end of the march, they declared their full satisfaction and support for the Islamic Emirate," Twitter post, January 20, 2022.

10 Muhammad Jalal [@MJalal313] (no official title), "Afghan women are rejecting those who are trying to represent them in Europe and elsewhere," Twitter post, February 5, 2022.

Islamic law. Women are the ones who raised and nurtured generations, men, heroes and philosophers.[11]

Western Concerns About Women's Rights Are Insincere

Attacks against U.S. and European involvement in Afghanistan with regard to women's rights are common. These attacks take several angles, including accusations that the U.S. campaign for women's rights was a pretext for imperialism and anti-Islamic agendas, videos of women being mistreated in Western countries, and women's rights abuses under the U.S.-backed Ghani regime. The following tweets are characteristic of this theme:

> The main pain of some is that projects that ridicule religious and national values and divisive programs are no longer applicable in Afghanistan. Human rights and freedom of expression are the only excuse![12]

> Today is International Women's Day. One of the major grievances of the rulers over the last 20 years has been the use of women as tools. The criterion for foreign donors to increase funding to government agencies was that the more women they hired, the more dollars they would pay. For the past 20 years, hiring women to attract dollars has become a tradition.[13]

Women's Rights Should Be Rooted in Islam

The IEA media team consistently asserts that women's rights in Afghanistan should align with Islamic principles. Tweets have focused on the hijab and on gender-segregated schools. For example, many tweets chastised the women taking part in the January 2022 protests against the countrywide hijab mandate. The following tweets are characteristic of this theme:

---

[11] Abdul Qahar Balkhi [@QaharBalhki] (MoFA Spokesperson), "Faryab Province: A peaceful women's march: We do not want democracy. We want our rights that Islam has given us. A group of immodest women do not represent Afghan women. Women's right to be educated. And to know. And to operate in the light of Islamic law. Women are the ones who raised and nurtured generations, men, heroes and philosophers," Twitter post, September 14, 2021.

[12] Inamullah Samangani [@HabibiSamangani], Deputy Spokesperson, "The main pain of some is that projects that ridicule religious and national values and divisive programs are no longer applicable in Afghanistan. Human rights and freedom of expression are the only excuse," Twitter post, February 4, 2022.

[13] Ahmadullah Muttaqi [@Ahmadmuttaqi01] (Assistant Chief of Staff to the Prime Minister, Deputy Director General Public and Strategic Affairs Office), "Today is International Women's Day. One of the major grievances of the rulers over the last 20 years has been the use of women as tools. The criterion for foreign donors to increase funding to government agencies was that the more women they hired, the more dollars they would pay. For the past 20 years, hiring women to attract dollars has become a tradition," Twitter post, March 8, 2022.

And someone has to tell you how you can say what this girl wants with your Ban Hijab slogan? Whatever secularism may be, but from the point of view of you fake secularists, only means enmity with Islam. As its examples can be seen in different parts of the world.[14]

Our sisters are sure that their schools will start. The Islamic Emirate is trying to build a mechanism that is in accordance with Islamic principles and national interests. Then all our sisters' schools and educational centers will start.[15]

Women's rights are the same rights as Islam.[16]

Reports of Women's Rights Abuses Are False

Taliban officials deny the validity of external reports of women's rights abuses. These tweets often claim that the sources for these reports are either fictitious or misinterpreted, all in the name of anti-Taliban propaganda. Media officials usually tweet these statements immediately following major news articles or human rights reports about Afghanistan. The following tweets are characteristic of this theme:

The claim of the Human Rights Organization that the Islamic Emirate does not address women's rights is mere propaganda and negative propaganda. The Islamic Emirate is committed to the rights of all citizens of the country, both men and women, and their right to work and education is protected within the framework of Islamic law.[17]

Women are working in Afghanistan. The reason that no one is showing that because in Afghan culture & society almost all women are against someone taking their pictures &

---

[14] Inamullah Samangani [@HabibiSamangani] (Deputy Spokesperson for the IEA), "And someone has to tell you how you can say what this girl wants with your Ban Hijab slogan? Whatever secularism may be, but from the point of view of you fake secularists, only means enmity with Islam. As its examples can be seen in different parts of the world," Twitter post, February 9, 2022.

[15] Bilal Karimi [@BilalKarimi21] (Deputy Spokesperson for the IEA), "Our sisters are sure that their schools will start. The Islamic Emirate is trying to build a mechanism that is in accordance with Islamic principles and national interests. Then all our sisters' schools and educational centers will start," Twitter post, November 25, 2021.

[16] Ahmadullah Muttaqi [@Ahmadmuttaqi01] (Assistant Chief of Staff to the Prime Minister, Deputy Director General Public and Strategic Affairs Office), "Women's rights are the same rights as Islam," Twitter post, March 8, 2021.

[17] Inamullah Samangani [@HabibiSamangani] (Deputy Spokesperson for the IEA), "The claim of the Human Rights Organization that the Islamic Emirate does not address women's rights is mere propaganda and negative propaganda. The Islamic Emirate is committed to the rights of all citizens of the country, both men and women, and their right to work and education is protected within the framework of Islamic law," Twitter post, December 27, 2021.

videos to make news out of it or post them on social media. Everyone, particularly westerners should understand.[18]

Afghan Women's Rights and Well-Being Require Foreign Assistance

The Taliban have linked the well-being and advancement of Afghan women to the amount of foreign aid given to the IEA. Officials have repeatedly asked for Taliban assets to be unfrozen, argued that a lack of jobs and education for women are because of a shortage of funds, and applauded instances of foreign aid. The following tweets are characteristic of this theme:

> China praises the positive actions of the Islamic Emirate and it is important to establish an inclusive government and a positive change in the lives of Afghan women and children. The Chinese Foreign Minister said that his country has provided urgent assistance to the people of Afghanistan and we are focusing on areas where urgent assistance is needed.[19]

> A number of women protested in #Kabul today against the country's frozen assets. The protesters demanded the release of frozen assets and the recognition of a new government.[20]

> The Islamic Emirate welcomes and appreciates the initiative of Germany and the Netherlands to pay the salaries of all male and female employees in the education and health sectors.[21]

Key Sources of Women's Rights Messaging

The Taliban officials and accounts that RAND identified for this study did not all discuss women's issues with the same frequency. Rather, a handful of accounts produced most of the tweets for this topic. Table 5.1 lists these top accounts.

---

[18] Muhammad Jalal [@MJalal313] (no official title), "Women are working in Afghanistan. The reason that no one is showing that because in Afghan culture & society almost all women are against someone taking their pictures & videos to make news out of it or post them on social media. Everyone, particularly westerners should understand," Twitter post, March 17, 2022.

[19] Inamullah Samangani [@HabibiSamangani], Deputy Spokesperson, "China praises the positive actions of the Islamic Emirate and it is important to establish an inclusive government and a positive change in the lives of Afghan women and children. The Chinese Foreign Minister said that his country has provided urgent assistance to the people of Afghanistan and we are focusing on areas where urgent assistance is needed," Twitter post, March 24, 2022.

[20] Al Emarah English [@Alemarahenglish] (official Taliban English Twitter account), "A number of women protested in #Kabul today against the country's frozen assets. The protesters demanded the release of frozen assets and the recognition of a new government," Twitter post, January 26, 2022.

[21] Zabihullah Mujahid [@Zabehulah_M33] (Deputy Minister of Culture and Information), "The Islamic Emirate welcomes and appreciates the initiative of Germany and the Netherlands to pay the salaries of all male and female employees in the education and health sectors," Twitter post, November 19, 2021.

TABLE 5.1
Top Producers of Women's Issues Messaging

Name	Title	Language by Frequency from Highest to Lowest
Mohammad Naeem Wardak	Spokesman of the Political Office	1. Arabic 2. Dari/Farsi 3. Pashto 4. English
Muhammad Jalal	Close to Taliban, but no official title	1. English 2. Pashto 3. Dari/Farsi
Al Emarah English	English version of the official Taliban publication	1. English
Inamullah Samangani	Deputy Spokesperson for the IEA	1. Dari/Farsi 2. English
Ahmadullah Wasiq	RTA Director	1. Pashto 2. Dari/Farsi 3. English

TABLE 5.2
Women's Issues Messaging by Language

Theme	Pashto and Dari/Farsi Emphasis	Arabic Emphasis	English Emphasis
The IEA supports women's education and workforce participation	High	Low	High
The IEA protects women's rights	High	Medium	High
Afghan women are happy with the new regime	Medium	High	Medium
Western concerns about women's rights are insincere	Low	High	Low
Women's rights should be rooted in Islam	High	High	Low
Reports of women's rights abuses are false	Medium	Low	Medium
Afghan women's rights and well-being require foreign assistance	High	Medium	High

Women's Rights Messaging by Language

Tweets in each language group discussed a different combination of the seven identified themes (Table 5.2). These are our findings by language group.

Arabic

The most-prominent themes in Arabic-language messaging are Afghan women's support for the regime, Western insincerity regarding women's issues, and the importance of having women's rights align with Islamic values. Frequent content included women marching in favor of hijab laws; footage of women being mistreated in Western countries; and support for gender-segregated schools. Afghan women's need for outside aid and the IEA's commitment to protecting women's rights had moderate prominence. Discussions about the Taliban's support for women's education and workforce participation, along with rejections of reports about women's rights abuses, were notably absent in Arabic tweets.

Pashto and Dari/Farsi

Four themes appeared frequently in Pashto and Dari/Farsi tweets: Taliban support for women's education and workforce participation; IEA support for equal rights for all women; the necessity for women's rights to conform to traditional interpretations of Islam; and the need for foreign aid to support women's advancement and well-being. Tweets gave special focus to the opening of gender-segregated schools and the Supreme Leader's declaration on women's rights. Content related to Afghan women's satisfaction with the regime and refutations of reports of women's rights abuses had a moderate presence. Taliban officials did not devote much Pashto and Dari/Farsi messaging to attacking Western approaches to women's rights.

English

The IEA's support for women's rights and women's education and employment featured relatively heavily in English-language messaging. Of note, compared with other languages, was the Taliban's celebration of International Women's Day to emphasize its support of women's rights. Afghan women's dependence on international aid for their advancement and well-being was another prominent theme. Media officials gave moderate focus to Afghan women's satisfaction with the regime and refutations of reports of women's rights abuses. They devoted limited focus to promoting Islamic women's rights and critiquing Western women's rights movements.

CHAPTER 6

Observations and Recommendations

This study was initiated to explore a significant question concerning Taliban use of social media communications: Are there differences in material communicated in the regional languages and between those languages and English? Our hypothesis was that differences in messaging between the languages could be identified to provide insight into current and future Taliban actions, especially those that might affect U.S. and Western interests—and, in doing so, help policymakers develop appropriate actions and responses.

We reviewed thousands of tweets, focusing most of our attention on several themes of interest to U.S. policymakers: government structure, economic and trade relations, relationships with neighboring countries and with other militant groups, and the status of women and education in Afghan society.

Taliban economic messaging in all languages is largely focused on the need for economic support to fix the broken economy. The messaging generally identifies what the government is doing to build the economy, but it continues to place blame on the United States and the prior regime for the broken economy. Messaging also is consistently critical of the United States for not providing support; this is done at least partly to deflect domestic attention from the slow pace of economic improvement.

Social media messaging on relationships with neighboring countries similarly communicated positive and critical content. Taliban focus was almost completely on economic ties, although connections with and the mention of Tajikistan was noticeably absent.

Regarding relationships with militant groups, the analysis noted the Taliban's adversarial relationship with ISKP. Not surprisingly, the Taliban are silent about regional and international foreign terrorist groups, such as IMU, the Eastern Turkistan Islamic Movement, Jaish-i-Mohammed, Jamaat Ansarullah, Lashkar-e-Tayyiba, and al Qaeda. The Taliban do not acknowledge relationships with these organizations and would likely express opposition to them in any public statements, including those on social media.

Regarding education and the status of women, there was no instance in which posts in one language directly contradicted those in another. However, thematic analysis across languages revealed differences. The most-notable differences were Arabic messaging's relatively low emphasis on the Taliban's support for women's education and workforce participation and relatively high emphasis on Western nations' insincere commitment to women's rights, and English messaging's relatively low emphasis on aligning women's rights with Islamic values. Messaging across the language groups was mixed and appeared to be tailored to the target audience: Taliban messaging in English was very supportive of women and girls in the workforce and education; tweets in other languages contradicted these supportive posts.

Analyzing the Taliban's use of social media from a strategic perspective, our research suggests that there is no overall strategy behind social media use. There is no strong evidence of an effort to systematically shape policy messaging or to systematically influence a target audience through repeated, tailored messaging. However, the Taliban are tailoring content on individual policy issues to specific groups. As noted, Taliban messaging on some issues was common across languages, but the content was distinctly different in other cases, reflecting an interest in tailoring material for a specific audience without regard to contradictions in messaging. One particularly notable result of our analysis was the gap between Taliban messaging and their actions. The best example of this conflict is the domestic messaging on women's rights and education and the messaging circulated in English to U.S. and Western governments. Domestic messaging in local languages on women's rights in Afghanistan emphasizes that these rights should align with Islamic principles, that women should have access to education and participate in the workforce, and that external criticisms of IEA policies toward women are based on falsehoods. By contrast, English-language messaging featured support for broadening women's education but provided only limited discussion of promoting Islamic women's rights and critiquing Western women's rights movements.

The study concept we followed—capturing, comparing, and analyzing Taliban social media commentary in key languages—has, in our view, demonstrated an ability to identify Taliban leadership policies, to highlight efforts to manage messaging to different constituencies, and to expose potentially conflicting views between and among leaders. We judge that sustained monitoring of Taliban messaging in the key languages of the region and in English will continue to provide insights on the evolution of Taliban policies and policy leaders. These insights are potentially extremely useful for U.S. and Western policymakers seeking to influence Taliban policies.

For Future Study

Our study's tight focus on thematic material in Taliban tweets and the narrow period we observed excluded any examination of other potentially important issues. The issues highlighted here represent a set of follow-on research topics that would continue to inform U.S. and Western policy deliberation toward the Taliban.

Continue to monitor Taliban messaging in the regional languages, particularly noting any differences in tone and content among the languages. The tight timeline of this study precluded any ability to assess whether Taliban use of social media was evolving from simple transactional use into a more strategic effort. We were also unable to conduct deeper analysis of the content of the tweets to identify potential conflicts between Taliban leaders or a more-detailed exploration of the gaps between Taliban statements and their actions. Continual review of the Taliban social media presence, including capabilities and access that are either foreclosed or not used at all, will provide U.S. and Western policymakers with a gauge of internal Taliban thinking on issues and potentially highlight potential points of convergence with U.S. and Western goals for Afghanistan.

Consider expanding the analysis of differences in messaging between English and regional languages to encompass other countries (such as Pakistan, India, and Iran) and militant groups (such as ISKP) to broaden understanding of the wider regional situation. Our initial review of available social media revealed that Pashto, Dari/Farsi, Arabic, Urdu, and English were the richest sources of data. We did not review social media in other languages that had lower density data, but the methodology used in this study might provide similar insights regarding important but lower-priority areas.

Consider applying a similar approach in monitoring the anti-Taliban (former Northern Alliance) groups and their military activities in the north and among former national security forces trained by NATO. Monitoring these groups' social media activities could provide similar insights into their plans and operations and provide U.S. and Western policymakers information that would otherwise be unavailable.

Consider establishing a full-time team of native linguists and analysts with area expertise to continue to monitor Taliban messaging and provide in-depth, culturally sensitive insight into longer-term Taliban goals and actions and leadership dynamics within the group. Close monitoring by native linguists of these social media activities could provide policymakers with a nuanced cultural appreciation of events that, in turn, could provide insights on domestic and regional developments. This stream of analysis could supplement traditional analysis based on diplomatic contacts and international press coverage and so provide a useful validation mechanism. It could also provide insights into early efforts by the Taliban to refocus priorities that would almost certainly create additional domestic and regional security uncertainties.

Abbreviations

IEA	Islamic Emirate of Afghanistan
IMU	Islamic Movement of Uzbekistan
ISKP	Islamic State Khorasan Province
MoFA	Ministry of Foreign Affairs
NATO	North Atlantic Treaty Organization
RTA	Radio Television Afghanistan
TAPI	Turkmenistan Afghanistan Pakistan India Gas Pipeline
TIP	Turkistan Islamic Party
TTP	Tehrik-e-Taliban Pakistan
UN	United Nations

References

The authors of this report provided the translations of bibliographic details for the non-English sources included in this report. The original rendering is not provided. Links are provided for posts that were still available online at the time of publication.

Afghan Biographies, "Ishaqzai, Abdul Hakim Mawlawi Sheikh," webpage, July 13, 2022. As of August 16, 2022:
http://www.afghan-bios.info/index.php?option=com_afghanbios&id=4698&task=view&total=699&start=272&Itemid=2

"Afghan Taliban Announce Successor to Mullah Mansour," BBC News, May 25, 2016.

Afghanistan International, "The Taliban Are Playing Double Games," video, April 13, 2022. As of August 16, 2022:
https://www.afintl.com/video/ott_6f9eff1a8b6b4c818bb40c831b026a58

Ahmadi, Qari Yousaf [@QyAhmadi21] (Spokesman of the Islamic Emirate of Afghanistan), "Information from the Ministry of Public Health of the Islamic Emirate of Afghanistan: The Ministry of Public Health of the Islamic Emirate informs all female employees to attend their duties regularly in the center and provinces. There is no impediment from the Islamic Emirate to carrying out their work," Twitter post, August 27, 2021. As of June 6, 2022:
https://twitter.com/QyAhmadi21/status/1431278131848044553

Ahmed, Hafiz Zia [@HafizZiaAhmad1] (Deputy Spokesman and Assistant Director of Public Relations, Ministry of Foreign Affairs), "The two sides held detailed discussions on current political & economic situation in Afghanistan. Both sides agreed that the Afghan Central Bank's USD $3.5 billion unfrozen assets from the US bank shall in no circumstances be given to charity organizations. The Afghan side," Twitter post, March 11, 2022. As of June 13, 2022:
https://twitter.com/HafizZiaAhmad1/status/1502357900353839104

Ahmed, Hafiz Zia [@HafizZiaAhmad1] (Deputy Spokesman and Assistant Director of Public Relations, Ministry of Foreign Affairs), "Today, Russian President's Special Envoy Zamir Kabulov led a delegation to Kabul and met with Foreign Minister Maulvi Amir Khan Mottaki. The meeting focused on strengthening political, economic, transit and regional ties. Mr. Kabulov said the policy of the Islamic Emirate is balanced, regional and global," Twitter post, March 24, 2022. As of June 13, 2022:
https://twitter.com/HafizZiaAhmad1/status/1506950166036852742

Al Emarah English [@Alemarahenglish] (official Taliban English Twitter account), "Minister of Mines at the AfG Economic Conference: Illegal mining has been stopped across the country. Dozens of small mines are being contracted out to domestic investors for legal extraction. Many large mines that require high capita," Twitter post, January 19, 2022.

Al Emarah English [@Alemarahenglish] (official Taliban English Twitter account), "A number of women protested in #Kabul today against the country's frozen assets. The protesters demanded the release of frozen assets and the recognition of a new government," Twitter post, January 26, 2022. As of June 6, 2022:
https://twitter.com/Alemarahenglish/status/1486304585593597960

"Allegations of Afghan Interference in Kazakhstan's Insecurity Are Unfounded," Voice of Jihad (Al Emarah English), February 2, 2022. As of June 13, 2022:
https://alemarahenglish.af/?p=49896

Arg Presidential Palace [@ARG_1880] (official Twitter account), "Amir-ul-Momineen Sheikh Maulvi Hibatullah Akhundzada's instructions to the directors of public institutions," Twitter post, December 18, 2021.

Arg Presidential Palace [@ARG_1880] (official Twitter account), "Mawlawi Abdul Salam Hanafi: We are fully committed to supporting the humanitarian assistance of the UN and other donor organizations. we assure the full security, access, flexible policies, and enabled environment for donor community in continuing their humanitarian assistance," Twitter post, January 18, 2022. As of August 16, 2022:
https://www.afintl.com/video/ott_6f9eff1a8b6b4c818bb40c831b026a58

Azizi, Atiqullah [@at_azizi1] (Deputy Minister for Culture and Arts), "In Afghanistan, wherever ISIS attacks, corrupt officials of the previous regime strongly support them. Bloody, America and Europe have turned a blind eye to it. Such support is neither terrorism nor fundamentalism? !!!," Twitter post, October 23, 2021. As of June 13, 2022:
https://twitter.com/at_azizi1/status/1451967997648596994

Bakhtar News Agency [@BakhtarNA] (state news agency), "The Da Afghanistan Bank, central bank, authorities say that as part of the humanitarian aid, 32 million US dollars arrived in Kabul and deposited to the Afghanistan International Bank," Twitter post, March 20, 2022. As of June 13, 2022:
https://twitter.com/BakhtarNA/status/1505505942741430282?s=20&t=wRXwOiKoSBN9AY mGo_IREg

Bakhtar News Agency [@BakhtarNA] (state news agency), "Mullah Abdul Ghani Barather, the First Deputy Prime Minister, traveled to the Balkh province to attend the inauguration ceremony of the Qosh-Tipa canal. Several other senior officials have also traveled for this purpose. This canal has a length of 285 kilometers, width 100 meters, and depth 805 meters," Twitter posts, March 29, 2022. As of June 13, 2022:
https://twitter.com/BakhtarNA/status/1508805389592277003
https://twitter.com/BakhtarNA/status/1508805490893107202

Bakhtar News Agency [@BakhtarNA] (state news agency), "Food items were distributed to more than 7500 families affected by war and drought in the Aqcha district of the Juzjan province. Mawlawi M. Zarif Fayez, the Juzjan Economic Director, said the assistance packages included flour, cooking oil, and peas sponsored by WFP," Twitter post, April 13, 2022. As of June 13, 2022:
https://twitter.com/BakhtarNA/status/1514201697987485702

Balkhi, Abdul Qahar [@QaharBalhki] (MoFA Spokesperson), "Faryab Province: A peaceful women's march: We do not want democracy. We want our rights that Islam has given us. A group of immodest women do not represent Afghan women. Women's right to be educated. And to know. And to operate in the light of Islamic law. Women are the ones who raised and nurtured generations, men, heroes and philosophers," Twitter post, September 14, 2021. As of June 6, 2022:
https://twitter.com/IeaOffice/status/1437804238487400454

Balkhi, Abdul Qahar [@QaharBalkhi] (MoFA Spokesperson), "Remarks by MoFA spokesperson regarding recent US action: MoFA of the IEA welcomes recent decision by US Treasury Department allowing US gov agencies along with international and nongovernmental organizations & banks to facilitate flow of food & medicine to the IEA," Twitter post, September 25, 2021. As of June 13, 2022:
https://twitter.com/QaharBalkhi/status/1441809931083128834

Balkhi, Abdul Qahar [@QaharBalkhi] (MoFA Spokesperson), "Transcript of Speech by the Acting Foreign Minister, H.E. Mawlawi Amir Khan Muttaqi," Twitter post, September 30, 2021. As of June 13, 2022:
https://twitter.com/QaharBalkhi/status/1443648415909490688

Balkhi, Abdul Qahar [@QaharBalkhi] (MoFA Spokesperson), "Transcript of Speech by the Deputy Prime Minister, His Excellency Mullah Abdul Ghani Baradar," Twitter post, September 30, 2021. As of June 13, 2022:
https://twitter.com/QaharBalkhi/status/1443640034087813124

Balkhi, Abdul Qahar [@QaharBalkhi] (MoFA Spokesperson), "Two-day dialogue between delegations of the Islamic Emirate and USA in Doha," Twitter post, October 10, 2021. As of June 13, 2022:
https://twitter.com/QaharBalkhi/status/1447308774495113216

Balkhi, Abdul Qahar [@QaharBalkhi] (MoFA Spokesperson), "On Friday a senior IEA delegation led by Foreign Minister Mawlawi Amir Khan Muttaqi met Yunus Sezer, head of Turkey's Disaster & Emergency Management Presidency & accompanying delegation. Mr. Sezer said they were ready to help Afghanistan in training & humanitarian assistance," Twitter post, October 17, 2021. As of June 13, 2022:
https://twitter.com/QaharBalkhi/status/1449776617618911232

Balkhi, Abdul Qahar [@QaharBalkhi] (MoFA Spokesperson), "her voice for the resumption of banking system of Afghanistan. Expressing gratitude to their efforts, Minister Muttaqi assured that the Islamic Emirate was fully prepared to deliver humanitarian aid," Twitter post, October 23, 2021. As of June 13, 2022:
https://twitter.com/QaharBalkhi/status/1451949933511204867

Balkhi, Abdul Qahar [@QaharBalkhi] (MoFA Spokesperson), "IEA delegation assured them about security and emphasized that Afghanistan seeks positive relations with the world, similarly urging transparent and urgent health assistance to Afghanistan. In the end, both sides agreed to continue such dialogue," Twitter post, November 28, 2021. As of June 13, 2022:
https://twitter.com/QaharBalkhi/status/1465029849840558080

Balkhi, Abdul Qahar [@QaharBalkhi] (MoFA Spokesperson), "IEA Foreign Minister H.E. Mawlawi Amir Khan Muttaqi and accompanying delegation met with the UK Charge d'Affaires for Afghanistan Mission Mr. Martin Longden & discussed the humanitarian, economic & security situation of Afghanistan," Twitter post, November 28, 2021. As of June 13, 2022:
https://twitter.com/QaharBalkhi/status/1465029724191834114

Balkhi, Abdul Qahar [@QaharBalkhi] (MoFA Spokesperson), "Afghan Foreign Minister Mawlawi Amir Khan Muttaqi met this evening with Uzbek delegation led by Minister of Transport Mr. Ilkhom Makhkamov. The meeting focused on constructive discussions on bilateral relation, the Uzbekistan delegation assured Minister Muttaqi of commencing," Twitter post, December 18, 2021. As of June 13, 2022:
https://twitter.com/QaharBalkhi/status/1472247988827291649

Balkhi, Abdul Qahar [@QaharBalkhi] (MoFA Spokesperson), "Uzbekistan-Afghanistan-Pakistan railway in the coming spring, saying Afghans will be trained by Uzbekistan on railway operation," Twitter post, December 18, 2021. As of June 13, 2022:
https://twitter.com/QaharBalkhi/status/1472247988827291649

Balkhi, Abdul Qahar [@QaharBalkhi] (MoFA Spokesperson), "IEA FM Mawlawi Amir Khan Muttaqi met Iranian FM Mr. Hossein Amir Abdollahian during the extraordinary OIC session on Afghanistan. The two sides talked on current security and humanitarian situation as well as trade, economy, and political relations between the two countries," Twitter post, December 19, 2021. As of June 13, 2022:
https://twitter.com/QaharBalkhi/status/1472556149304172549

Balkhi, Abdul Qahar [@QaharBalkhi] (MoFA Spokesperson), "The Turkish Foreign Minister said Turkish companies and traders would visit Afghanistan. Minister Muttaqi said security and facilities would be provided for Turkish traders and humanitarian organization and all impediments would be removed," Twitter post, December 19, 2021. As of June 13, 2022:
https://twitter.com/QaharBalkhi/status/1472535163875868676

Balkhi, Abdul Qahar [@QaharBalkhi] (MoFA Spokesperson), "The freezing of foreign exchange reserves and the associated banking crises deteriorates the humanitarian crisis further," Twitter post, December 21, 2021. As of June 13, 2022:
https://twitter.com/BakhtarNA/status/1514201697987485702

Balkhi, Abdul Qahar [@QaharBalkhi] (MoFA Spokesperson), "We appreciate step by 48 US Congress members who, following our open letter to the Congress, have recommended President Joe Biden (@POTUS) lift sanctions imposed on Afghanistan & unfreeze its Central Bank reserves," Twitter post, December 21, 2021. As of June 13, 2022:
https://twitter.com/QaharBalkhi/status/1473561943684304898

Balkhi, Abdul Qahar [@QaharBalkhi] (MoFA Spokesperson), "Following his arrival in Kabul today, UAE Minister of Federal Authority for Identity, Citizenship, Customs & Ports Security Mr. Ali Mohammed bin Hammad Al Shamsi discussed issues of importance with IEA Foreign Minister Mawlawi Amir Khan Muttaqi," Twitter post, December 28, 2021. As of June 13, 2022:
https://twitter.com/QaharBalkhi/status/1476014770042155010

Balkhi, Abdul Qahar [@QaharBalkhi] (MoFA Spokesperson), "Recently a few of incidents have taken place along Durand line between Afghanistan and Pakistan that have given rise to the need for authorities of the two sides to address the problem," Twitter post, January 4, 2022. As of June 13, 2022:
https://twitter.com/QaharBalkhi/status/1478396845013491718

Balkhi, Abdul Qahar [@QaharBalkhi] (MoFA Spokesperson), "The Ministry of Foreign Affairs of IEA strongly rejects remarks by President @POTUS asserting Afghanistan is incapable of unity," Twitter post, January 20, 2022. As of June 13, 2022:
https://twitter.com/QaharBalkhi/status/1484422427786440704

Balkhi, Abdul Qahar [@QaharBalkhi] (MoFA Spokesperson), "Also the Foreign Minister requested the Chinese Ambassador to expedite the work on mining projects that China has invested in," Twitter post, February 3, 2022. As of June 13, 2022:
https://twitter.com/QaharBalkhi/status/1489238091948871686

Balkhi, Abdul Qahar [@QaharBalkhi] (MoFA Spokesperson), "said it was not true, adding Afghans consider such rhetoric concerning, provocative, & direct incitement & campaign for subversive groups. Minister Muttaqi said the Afghan government will allow none to use the Afghan soil against another," Twitter post, February 22, 2022. As of June 13, 2022:
https://twitter.com/QaharBalkhi/status/1496197397046870029

Balkhi, Abdul Qahar [@QaharBalkhi] (MoFA Spokesperson), "Afghan Foreign Minister Mawlawi Amir Khan Muttaqi welcomed Chinese Foreign Minister Wang Yi to Kabul in a special visit to Afghanistan. The Foreign Ministers met in Storai Palace-MoFA to discuss political, economic & transit issues, air corridor, dried fruit export, educational," Twitter post, March 24, 2022. As of June 13, 2022:
https://twitter.com/QaharBalkhi/status/1506940252178558980

Balkhi, Abdul Qahar [@QaharBalkhi] (MoFA Spokesperson), "Minister Muttaqi said the new government is focused on regional security & connectivity, & wants Afghanistan, as the heart of Asia, to play an important role in strengthening transit, trade, industry & economy among the countries of the region," Twitter post, March 24, 2022. As of June 13, 2022:
https://twitter.com/QaharBalkhi/status/1506972524294377473

Balkhi, Abdul Qahar [@QaharBalkhi] (MoFA Spokesperson), "The Ministry of Foreign Affairs of IEA is saddened to learn about the rocket attacks on oil facilities and civilian targets in Jeddah-KSA, and condemns attacks on civilian targets. IEA believes that such acts threaten regional peace and stability and should be ceased," Twitter post, March 26, 2022. As of June 13, 2022:
https://twitter.com/QaharBalkhi/status/1507705354573942784

Benard, Cheryl, Seth G. Jones, Olga Oliker, Cathryn Quantic Thurston, Brooke Stearns Lawson, and Kristen Cordell, *Women and Nation-Building*, RAND Corporation, MG-579-IMEY/CMEPP, 2008. As of June 6, 2022:
https://www.rand.org/pubs/monographs/MG579.html

Bezhan, Frud, "The Rise of Mullah Yaqoob, the Taliban's New Military Chief," Radio Free Europe/Radio Liberty, August 27, 2020.

Byrd, William, "Taliban Are Collecting Revenue—But How Are They Spending It?" Institute of Peace, February 2, 2022.

Central Intelligence Agency, "Afghanistan," World Factbook, May 31, 2022.

DataReportal, "Digital 2022: Afghanistan," webpage, February 15, 2022. As of June 12, 2022:
https://datareportal.com/reports/digital-2022-afghanistan

Dreazen, Yochi, "The Taliban's New Number 2 Is a 'Mix of Tony Soprano and Che Guevara,'" *Foreign Policy*, July 31, 2015.

"Taliban Has Doubled Coal Prices," DW, July 7, 2022. As of January 23, 2023:
https://p.dw.com/p/4DnTS

Federal Bureau of Investigation, "Most Wanted: Sirajuddin Haqqani," webpage, undated. As of April 25, 2022:
https://www.fbi.gov/wanted/terrorinfo/sirajuddin-haqqani

First Deputy Prime Minister Office [@FDPM_AFG] (official Twitter account of the Office of Mullah Abdul Ghani Barader), "and invest in mining, creating economic zones and energy production. The Chinese Foreign Minister added that they will soon start extracting the Mes Aynak mine," Twitter post, March 24, 2022. As of June 13, 2022:
https://twitter.com/FDPM_AFG/status/1506942425016766467

General Directorate of Intelligence [@GDI1415], "Dear Countrymen! With the divine assistance of almighty Allah (SWT), the unyielding support of the courageous Afghan nation, and the unbelievable sacrifices of our holy warriors, our beloved country has once again been blessed with a truly Islamic government, and security that is steadily prevailing throughout the country. The leadership and staff at the General Directorate of Intelligence of the Islamic Emirate of Afghanistan consider protection of national interests and Islamic values as their religious duty and national obligation, and they proudly endure any tiresome of hardships to ensure the security of the countrymen. The General Directorate of Intelligence urges all the compatriots to assist us in establishing absolute security and foiling the nefarious plots of evil elements in a timely manner by urgently reporting any suspicious activity from individuals within your regions to local security forces, or to call us free toll number 1001. Regards, the General Directorate of Intelligence (GDI) 26/10/2021 Gregorian," Twitter post, October 26, 2021. As of June 13, 2022:
https://twitter.com/GDI1415/status/1453014378576498708

"The Guantánamo Docket," *New York Times*, May 18, 2021.

Haqqani, Abdul Hakim, *The Islamic Emirate and Its System*, Darul-Ulum Shariyah, 2022.

Haqqani, Anas [@AnasHaqqani313] (senior member of the Taliban), "Khalifa Sirajuddin Haqqani, the legendary leader, and one of the great leaders against the invading forces and the leader of the battle of freedom. He gave a new life to his heroism by appearing in front of the journalists' lenses. The defeat of NATO was the result of the loyalty, altruism and leadership of these leaders to the liberation battle," Twitter post, March 5, 2022.

"Hibatullah's Roots Were Non-Political and Reclusive," TOLOnews, May 2016.

Information TV, "Sher Mohammad Abbas Stanikzai' Speech in the West of Kabul," video, December 29, 2021.

Jalal, Muhammad [@MJalal313] (no official title), "Deputy Islamic Emirate and Minister of Interior His Excellency Khalifa Sirajuddin Haqqani, may God protect him. He who fought against the so-called superpower of our time and brought them to their knees in Afghanistan," Twitter post, November 8, 2021. As of June 13, 2022:
https://twitter.com/MJalal700/status/1457682392244707331

Jalal, Muhammad [@MJalal313] (no official title), "'Amr al-Ma'ruf does not mean that a veiled woman, whose whole body is covered except her face, should be covered under the name of Islam and half of her body should be exposed in the beating.' Sheikh Khalid Hanafi (Acting Minister)," Twitter post, November 17, 2021.

Jalal, Muhammad [@MJalal313] (no official title), "The Ministry of Interior has appointed anti-narcotics officials in all provinces. The Deputy Minister for Counter Narcotics has said that he has made recommendations to officials to stop the sale & purchase of narcotics & poppy cultivation in all provinces as soon as possible," Twitter post, December 11, 2021. As of June 5, 2022:
https://twitter.com/MJalal313/status/1489858362057666564

Jalal, Muhammad [@MJalal313] (no official title), "Siraj-ud-Din Haqqani, the true defender of Islamic religion," Twitter post, December 31, 2021.

Jalal, Muhammad [@MJalal313] (no official title), "The Amir of IEA, H.E Sheikh Hibatullah Akhundzada paid a visit to Herat province and gave necessary instructions to the officials there. This is the third province that he has visited. Previously he went to Farah province," Twitter post, January 12, 2022. As of June 5, 2022:
https://twitter.com/MJalal700/status/1457682392244707331

Jalal, Muhammad [@MJalal313] (no official title), "Update: Former head of Daesh- ISKP terrorist group, Aslam Farooqi has been killed in Northern Afghanistan. He was at large and involved in crimes against Afghanistan. #Peace," Twitter post, January 16, 2022. As of June 13, 2022:
https://twitter.com/MJalal313/status/1482792364653621250

Jalal, Muhammad [@MJalal313] (no official title), "Afghan women are rejecting those who are trying to represent them in Europe and elsewhere," Twitter post, February 5, 2022. As of June 6, 2022:
https://twitter.com/MJalal313/status/1489858362057666564

Jalal, Muhammad [@MJalal313] (no official title), "American governments terrorized Afghans for 20 years and now your boss @POTUS is stealing the money belonging to innocent Afghans," Twitter post, February 11, 2022. As on June 13, 2022:
https://twitter.com/MJalal313/status/1492330408117211136

Jalal, Muhammad [@MJalal313] (no official title), "Statement of MoFA, IEA on Russia-Ukraine conflict: The Islamic Emirate of Afghanistan, in line with its foreign policy of neutrality, calls on both sides of the conflict to resolve the crisis through dialogue and peaceful means. @QaharBalkhi," Twitter post, February 25, 2022. As of June 13, 2022:
https://twitter.com/MJalal313/status/1497127705808101402

Jalal, Muhammad [@MJalal313] (no official title), "The decision to open all schools in Afghanistan for all students is because the Islamic Emirate of Afghanistan wants education for all Afghans. Providing education to all Afghans is one of the important goals of IEA," Twitter post, March 3, 2022.

Jalal, Muhammad [@MJalal313] (no official title), "The message of H.E Khalifa Sirajuddin Haqqani was different than the way his image was portrayed in western media. He talked about respecting the sovereignty of our country, relations with international community. He talked about the safety & security of Afghans and Afghanistan," Twitter post, March 5, 2022. As of June 5, 2022:
https://twitter.com/MJalal700/status/1500049780201955331

Jalal, Muhammad [@MJalal313] (no official title), "H.E Khalifa Sirajuddin Haqqani: We will not allow anyone to use Afghan soil against others," Twitter post, March 6, 2022. As of June 13, 2022:
https://twitter.com/MJalal313/status/1500463384193552387

Jalal, Muhammad [@MJalal313] (no official title), "Picture is from Defense Minister's visit to the southern and western borders of the country," Twitter post, March 17, 2022. As of June 5, 2022:
https://twitter.com/MJalal700/status/1504475859737985025

Jalal, Muhammad [@MJalal313] (no official title), "Women are working in Afghanistan. The reason that no one is showing that because in Afghan culture & society almost all women are against someone taking their pictures & videos to make news out of it or post them on social media. Everyone, particularly westerners should understand," Twitter post, March 17, 2022. As of June 5, 2022:
https://twitter.com/MJalal700/status/1504359024095027200

Jalal, Muhammad [@MJalal313] (no official title), "GDI: We assure our nation, our neighbors and the world that Daesh is no longer a matter of concern in Afghanistan," Twitter post, March 20, 2022. As of June 13, 2022:
https://twitter.com/MJalal313/status/1505482073586417664

Jalal, Muhammad [@MJalal313] (no official title), "H.E Khalifa Sirajuddin Haqqani: We are committed in all our commitments. Afghanistan will no longer be a threat to any of its neighbors and other countries," Twitter post, March 24, 2022. As of June 13, 2022: https://twitter.com/MJalal313/status/1507043560788570114

Karimi, Bilal [@BilalKarimi21] (Deputy Spokesperson for the IEA), "Our sisters are sure that their schools will start. The Islamic Emirate is trying to build a mechanism that is in accordance with Islamic principles and national interests. Then all our sisters' schools and educational centers will start," Twitter post, November 25, 2021.

Karimi, Bilal [@BilalKarimi21] (Deputy Spokesperson for the IEA), "Zabihullah Mujahid: today, we are thrilled to roll out the Qosh-Tipa canal. With the completion of this project, we will not be dependent on grains from other countries, and public support is critical to the success of this project. The Islamic Emirate will use all the possible means at its disposal to complete this project," Twitter post, March 30, 2022. As of June 12, 2022: https://twitter.com/BilalKarimi21/status/1509065368815476738?s=20&t=C3lIt8 z5x-XclpgKLdnuMQ

Kaura, Vinay, "Tajikistan's Evolving Relations with the Taliban 2.0," Middle East Institute, December 1, 2021.

Khosty, Qari Saeed [@SaeedKhosty] (Ministry of Interior Public Relations and Press Director), "2718 drug addicts taken to hospitals in Kabul Over the past week, at least 2718 drug addicts were collected from various areas in Kabul & were sent to drug treatment centers in Kabul. According to IM Officials, collected addicts were taken to 14OPD cntrs & 14inpatient hospitals," Twitter post, November 20, 2022. As of June 5, 2022: https://twitter.com/saeedkhosty/status/1462012771005079557

Lee, Matthew, Nomaan Merchant, and Mike Balsamo, "CIA Drone Strike Kills al-Qaida Leader Ayman al-Zawahri in Afghanistan," *PBS News Hour*, August 1, 2022.

Marcolini, Barbara, Sanjar Sohail, and Alexander Stockton, "The Taliban Promised Them Amnesty. Then They Executed Them," *New York Times*, April 12, 2022.

Mashal, Mujib, and Taimoor Shah, "Taliban's New Leader, More Scholar Than Fighter, Is Slow to Impose Himself," *New York Times*, July 11, 2016.

"Maulvi Muhammad Yaqub Mujahid's Speech in the 9th Year Celebration of Amirul Momineen Mullah Muhammad Umar Mujahid," Radio Television Afghanistan (RTA) Pashto, 2022.

Meta, "Dangerous Individuals and Organizations," webpage, undated. As of August 20, 2022: https://transparency.fb.com/policies/community-standards/ dangerous-individuals-organizations/

Ministry of Agriculture, Irrigation and Livestock, "Programs," undated.

Ministry of National Defense, Afghanistan [@modafghanistan2] (official Twitter account), "A delegation from the Ministry of Defense met with the Deputy Chief of Mission of the Islamic Republic of Iran in Kabul," Twitter post, April 26, 2022. As of June 13, 2022: https://twitter.com/modafghanistan2/status/1518966404061540353

Ministry of Foreign Affairs—Afghanistan, "Biography Minister of Foreign Affairs of Afghanistan," webpage, undated. As of January 26, 2023: https://mfa.gov.af/en/minister-biography/

Ministry of Foreign Affairs—Afghanistan [@MoFA_Afg] (official account of the Ministry of Foreign Affairs of Afghanistan), "IEA Deputy Foreign Minister Alhaj Sher Mohammad Abbas Stanekzai met today with Mr. Wafa Khadzhiev, Deputy Foreign Minister of Turkmenistan. Mr. Khadzhiev expressed satisfaction with the overall security situation in Afghanistan, saying Turkmenistan would start work on TAPI," Twitter post, January 8, 2022. As of June 13, 2022: https://twitter.com/MoFA_Afg/status/1479819515592687619

Ministry of Foreign Affairs—Afghanistan [@MoFA_Afg] (official Twitter account), "A meeting between the Islamic Emirate delegation led by Acting Foreign Minister Mawlawi Amir Khan Muttaqi was held with Qatari officials from several civil service and education institutions. The meeting focused on humanitarian situation in Afghanistan, dynamics of higher," Twitter post, February 15, 2022. As on June 13, 2022: https://twitter.com/MoFA_Afg/status/1493820558861082625

Ministry of Foreign Affairs—Afghanistan [@MoFA_Afg] (official Twitter account), "education, capacity development, humanitarian aid, and providing scholarships to Afghan students. The Qatari officials pledged to assist in various areas, provide training opportunities for Afghan students, and deliver aid to Afghanistan through Qatari Red Crescent and," Twitter post, February 15, 2022. As on June 13, 2022: https://twitter.com/MoFA_Afg/status/1493820561276944388

Mujahid, Zabihullah [@Zabehulah_M33] (Deputy Minister of Culture and Information and Spokesperson), "1/3. Press Release of the Ministry of Foreign Affairs: As a result of the efforts, negotiations and communication of the Ministry of Foreign Affairs, the Islamic Emirate of Pakistan, in order to solve issues regarding the export of fresh fruit," Twitter post, September 25, 2021. As of June 13, 2022: https://twitter.com/Zabehulah_M33/status/1441840279603056640

Mujahid, Zabihullah [@Zabehulah_M33] (Deputy Minister of Culture and Information and Spokesperson), "has lifted 17% tax (Sale tax) that had been imposed some time back with effect from September 24, 2021. The Ministry of Foreign Affairs of the Islamic Emirate of Pakistan welcomes this step of the Islamic Republic of Pakistan," Twitter post, September 25, 2021. As of June 13, 2022: https://twitter.com/Zabehulah_M33/status/1441840282933215232

Mujahid, Zabihullah [@Zabehulah_M33] (Deputy Minister of Culture and Information and Spokesperson), "Qureshi: Immediate visas will be issued to Afghan patients in Torkham and will facilitate their travel," Twitter post, October 21, 2021. As of June 13, 2022: https://twitter.com/Zabehulah_M33/status/1451237702867816448

Mujahid, Zabihullah [@Zabehulah_M33] (Deputy Minister of Culture and Information and Spokesperson), "Important: An important meeting was held by the leaders of the Islamic Emirate with a delegation led by the Minister of Foreign Affairs of Turkmenistan Rashid Muradov at the Presidential Palace today. Economic, security, trade and humanitarian assistance were discussed at the meeting," Twitter post, October 30, 2021. As of June 13, 2022: https://twitter.com/Zabehulah_M33/status/1454365056045551617

Mujahid, Zabihullah [@Zabehulah_M33] (Deputy Minister of Culture and Information and Spokesperson), "Foreign Minister of Turkmenistan Muradov and his accompanying delegation had a special meeting with the Minister of Defense of the Islamic Emirate of Afghanistan Maulvi Mohammad Yaqub Mujahid and other officials at the Presidential Palace. The meeting discussed in detail the security situation on the border between the two countries and in particular the security measures for the TAPI project," Twitter post, October 30, 2021. As of June 13, 2022: https://twitter.com/Zabehulah_M33/status/1454632475892793345

Mujahid, Zabihullah [@Zabehulah_M33] (Deputy Minister of Culture and Information and Spokesperson), "4/6 and Afghanistan's economy improves somewhat. Maulvi Abdul Salam Hanafi thanked the WFP for its cooperation with Afghanistan in various fields and said, 'We are ready for any kind of cooperation with all the UN agencies and ensure their full security,'" Twitter post, November 7, 2021. As of June 13, 2022:
https://twitter.com/Zabehulah_M33/status/1457350800745893889

Mujahid, Zabihullah [@Zabehulah_M33] (Deputy Minister of Culture and Information), "The Islamic Emirate welcomes and appreciates the initiative of Germany and the Netherlands to pay the salaries of all male and female employees in the education and health sectors," Twitter post, November 19, 2021. As of June 13, 2022:
https://twitter.com/Zabehulah_M33/status/1461690246953934850

Mujahid, Zabihullah [@Zabehulah_M33] (Deputy Minister of Culture and Information and Spokesperson), "The incident between the Afghan and Iranian border guards in Nimroz was brought under control. The incident took place between the Afghan and Iranian border guards in Nimroz province," Twitter post, December 1, 2021. As of June 13, 2022:
https://twitter.com/Zabehulah_M33/status/1466110970414223361

Mujahid, Zabihullah [@Zabehulah_M33] (Deputy Minister of Culture and Information and Spokesperson), "Officials at Da Afghanistan Bank say, as part of series of humanitarian aid to #Afghanistan, $32 million in cash arrived in #Kabul today (Monday, February 21) and transferred to the Afghanistan International Bank (AIB)," Twitter post, February 21, 2022.

Mujahid, Zabihullah [@Zabehulah_M33] (Deputy Minister of Culture and Information and Spokesperson), "A local incident took place in Spin Boldak area of Kandahar with Pakistani guards. Unfortunately, the first shots were fired by Pakistani guards. Leaders on both sides have been informed and the situation is now under control. We will do a thorough investigation into why this happened," Twitter post, February 24, 2022. As of June 13, 2022:
https://twitter.com/Zabehulah_M33/status/1496854780223934471

Mujahid, Zabihullah [@Zabehulah_M33] (Deputy Minister of Culture and Information and Spokesperson), "As part of its assistance to Afghans, Pakistan has pledged 50,000 tonnes of wheat to Afghanistan, with the first batch of 3600 tonnes arrived in Nangarhar yesterday. We welcome this assistance from our brotherly country Pakistan and ask for more cooperation," Twitter post, March 1, 2022. As of June 13, 2022:
https://twitter.com/Zabehulah_M33/status/1498891736281665536

Mujahid, Zabihullah [@Zabehulah_M33] (Deputy Minister of Culture and Information and Spokesperson), "Sympathy: We condemn the bombing of a mosque in Peshawar, Pakistan. There is no justification for attacking civilians and worshipers. We express our deepest condolences to all the victims of the incident," Twitter post, March 4, 2022. As of June 13,2022:
https://twitter.com/Zabehulah_M33/status/1499802887672799235

Mujahid, Zabihullah [@Zabehulah_M33] (Central Spokesperson and Deputy Minister of Culture and Information), "IEA is fully committed to upholding of all the Sharia rights of the Afghan women. International Women's Day is an opportunity for our Afghan women to demand their legitimate rights. We will protect and defend the rights of our Afghan women, Insha Allah," Twitter post, March 8, 2022. As of June 15, 2022:
https://twitter.com/Zabehulah_M33/status/1501103977429143552

Mujahid, Zabihullah [@Zabehulah_M33] (Deputy Minister of Culture and Information and Spokesperson), "1/2- The Islamic Emirate of Afghanistan strongly condemns Pakistan's attacks on refugees in Khost and Kunar. IEA calls on the Pakistani side not to test the patience of Afghans on such issues and not repeat the same mistake again otherwise it will have bad consequences," Twitter post, April 16, 2022. As of June 13, 2022:
https://twitter.com/Zabehulah_M33/status/1515348809299202051

Muttaqi, Ahmadullah [@Ahmadmuttaqi01] (Assistant Chief of Staff to the Prime Minister, Deputy Director General Public and Strategic Affairs Office), "Women's rights are the same rights as Islam," Twitter post, March 8, 2022. As of June 15, 2022:
https://twitter.com/Ahmadmuttaqi01/status/1501127077973884932

Muttaqi, Ahmadullah [@Ahmadmuttaqi01] (Assistant Chief of Staff to the Prime Minister, Deputy Director General Public and Strategic Affairs Office), "By order of Amir al-mu'minin, Sheikh Hadith Maulvi Abdul Hakim Haqqani appointed as the head of Supreme Court, Sheikh Maulvi Mohammad Qasim Turkman as the first deputy, and Sheikh Maulvi Abdul Malik appointed as the second deputy," Twitter post, October 15, 2021.

Muttaqi, Ahmadullah [@Ahmadmuttaqi01] (Assistant Chief of Staff to the Prime Minister, Deputy Director General Public and Strategic Affairs Office), "Today is International Women's Day. One of the major grievances of the rulers over the last 20 years has been the use of women as tools. The criterion for foreign donors to increase funding to government agencies was that the more women they hired, the more dollars they would pay. For the past 20 years, hiring women to attract dollars has become a tradition," Twitter post, March 8, 2022.

Muttaqi, Ahmadullah [@Ahmadmuttaqi01] (Assistant Chief of Staff to the Prime Minister, Deputy Director General Public and Strategic Affairs Office), "The orders of Commander of the Faithful should be implemented 100 percent," Twitter post, March 23, 2022. As of June 15, 2022:
https://twitter.com/Ahmadmuttaqi01/status/1506559601021493251

National Counterterrorism Center, "Haqqani Network," webpage, undated. As of January 9, 2023:
https://www.dni.gov/nctc/groups/haqqani_network.html#:~:text=The%20Haqqani%20Network%20is%20a,renowned%20mujahedin%20commander%20Younis%20Khalis.

Nichols, Michelle, "U.N. Warns of 'Colossal' Collapse of Afghan Banking System," Reuters, November 22, 2021.

Nichols, Michelle, "U.N. Has Millions in Afghanistan Bank, but Cannot Use It," Reuters, February 3, 2022.

Office of First Deputy Prime Minister [@FDPM_AFG] (official Twitter account), "1/6: The First Deputy Prime Minister met with Esmatullah Agha, Special Representative of the President of Uzbekistan for Afghanistan! During the meeting, which took place between the two sides in the Citadel today, Tuesday, political and economic issues were discussed," Twitter post, November 30, 2021. As of June 13, 2022:
https://twitter.com/FDPM_AFG/status/1465661111538262018

Office of First Deputy Prime Minister [@FDPM_AFG] (official Twitter account), "5/6: He added that he would decide on helicopters and all military equipment transferred to their country in accordance with international law," Twitter post, November 30, 2021. As of June 13, 2022:
https://twitter.com/FDPM_AFG/status/1465661119939420175

Official Journal of the European Union, "2014/142/CFSP of 14 March 2014 implementing Decision 2011/486/CFSP concerning restrictive measures directed against certain individuals, groups, undertakings and entities in view of the situation in Afghanistan," Vol. 57, March 15, 2014.

"An Overview of the Activities of Violation of Citizens' Rights by the Ministry of Propagation of Virtue and the Prevention of Vice," *Etilaatroz Daily*, May 14, 2022.

Radio Television Afghanistan, "Sher Mohammad Abbas Stanikzai: We Have Always Turned Those Blind Who Had Bad Intent About Afghanistan," video, December 29, 2021. As of June 12, 2022:
https://www.youtube.com/watch?v=C31cb97uB10

Radio Television Afghanistan World, "Mawlawi Mohammad Yaqub Mujahid Acting Defense Minister Exclusive Interview with English Subtitles," video, February 15, 2022. As of June 13, 2022:
https://www.youtube.com/watch?v=rKUqmTbspB4&lc=UgzrUOgxo9D5ebB2uel4AaABAg

Rewards for Justice, "Sirajuddin Haqqani," webpage, undated. As of April 25, 2022:
https://rewardsforjustice.net/rewards/sirajuddin-haqqani

RTA—*See* Radio Television Afghanistan.

Samangani, Inamullah [@HabibiSamangani] (Deputy Spokesperson for the IEA), "Foreign Minister Maulvi Amir Khan Muttaqi, Trade and Finance Ministers, and their accompanying delegation met with Pakistani Prime Minister Imran Khan, cabinet members and other high-ranking officials. The Foreign Minister called for facilities to be provided at Torkham and Chaman Gates to facilitate the transit of Afghan goods 1/2," Twitter post, November 12, 2021. As of June 13, 2022:
https://twitter.com/HabibiSamangani/status/1459178820125270025

Samangani, Inamullah [@HabibiSamangani] (Deputy Spokesperson for the IEA), "The claim of the Human Rights Organization that the Islamic Emirate does not address women's rights is mere propaganda and negative propaganda. The Islamic Emirate is committed to the rights of all citizens of the country, both men and women, and their right to work and education is protected within the framework of Islamic law," Twitter post, December 27, 2021.

Samangani, Inamullah [@HabibiSamangani] (Deputy Spokesperson for the IEA), "Mullah Abdul Ghani Barader on RTA: We support investment in legal mining and export of minerals to create jobs and attain economic self-sufficiency," Twitter post, January 3, 2022. As of June 5, 2022:
https://twitter.com/HabibiSamangani/status/1478062977018474500?s=20&t=57xL9IRDIlgQqBVz58lbPA

Samangani, Inamullah [@HabibiSamangani] (Deputy Spokesperson for the IEA), "Ministry of Foreign Affairs: Reports of a dispute on the Afghan-Turkmen border are false. There is no problem between us and Turkmenistan's neighbor, and we want to have positive and constructive relations diplomatically with the friendly country of Turkmenistan and other countries of the world," Twitter post, January 4, 2022. As of June 13, 2022:
https://twitter.com/HabibiSamangani/status/1478355525645520898

Samangani, Inamullah [@HabibiSamangani] (Deputy Spokesperson for the IEA), "As a result of India's humanitarian assistance to the Afghans, six tons of various medicines arrived in Kabul from that country. India has already sent 500,000 doses of corona vaccine to Afghanistan. The Islamic Emirate of Afghanistan would like to thank India for its humanitarian assistance," Twitter post, January 7, 2022. As of June 13, 2022:
https://twitter.com/HabibiSamangani/status/1479373606325043202

Samangani, Inamullah [@HabibiSamangani] (Deputy Spokesperson for the IEA), "The claim of the President of Tajikistan that many camps have been set up in Afghanistan in the border areas with Tajikistan for destructive activities is not true and we reject it seriously. The Islamic Emirate assures all neighboring countries that our borders are safe and . . . 1/2," Twitter post, January 11, 2022. As of June 13, 2022:
https://twitter.com/HabibiSamangani/status/1480855305768554496

Samangani, Inamullah [@HabibiSamangani] (Deputy Spokesperson for the IEA), "There is no threat to any country from our territory, including Tajikistan. Some circles and fugitives and biased people transmit false information to neighboring countries and the world, which is never true. 2/2," Twitter post, January 11, 2022. As of June 15, 2022:
https://twitter.com/HabibiSamangani/status/1480855308549464067

Samangani, Inamullah [@HabibiSamangani] (Deputy Spokesperson for the IEA), "He assured that China would not interfere in Afghanistan's internal affairs. 'Based on the experience of the past 20 years, we call on the international community to allow Afghans to form their own governments and not to try to impose foreign ideas on Afghans,' he said. 2/4," Twitter post, January 15, 2022. As of June 13, 2022:
https://twitter.com/HabibiSamangani/status/1482400346949496843

Samangani, Inamullah [@HabibiSamangani] (Deputy Spokesperson for the IEA), "In today's economic conference, the representatives of 20 countries attended in person and the representative of other 40 countries virtually. The participants will discuss the economic issues of Afghanistan, particularly the banking problem and challenges facing the private sector, and recommend solutions," Twitter post, January 19, 2022. As of June 5, 2022:
https://twitter.com/HabibiSamangani/status/1483746360906113024

Samangani, Inamullah [@HabibiSamangani] (Deputy Spokesperson for the IEA), "The Ministry of Mining and Petroleum has opened the bidding process for ten small-size mining projects in Kandahar, Logar, and Zabul. The interested entities can submit their application within 15 working days and receive the other required documents after the initial screening," Twitter post, January 23, 2022. As of June 5, 2022:
https://twitter.com/HabibiSamangani/status/1485232098449444866

Samangani, Inamullah [@HabibiSamangani] (Deputy Spokesperson for the IEA), "2/2 The acting Minister of Economy thanked UN and the donor countries. And explained the new procedure approved by the IEA cabinet to coordinate humanitarian efforts, ensure transparency in aid distribution, and promised support for the humanitarian work," Twitter post, January 31, 2022. As of June 5, 2022:
https://twitter.com/HabibiSamangani/status/1488095240460320769?s=20&t=pVw-oktQ7BL83Udf-kb8CA

Samangani, Inamullah [@HabibiSamangani] (Deputy Spokesperson for the IEA), "The main pain of some is that projects that ridicule religious and national values and divisive programs are no longer applicable in Afghanistan. Human rights and freedom of expression are the only excuse," Twitter post, February 4, 2022. As of June 6, 2022:
https://twitter.com/HabibiSamangani/status/1489550707330101249

Samangani, Inamullah [@HabibiSamangani] (Deputy Spokesperson for the IEA), "Ministry of Foreign Affairs: The report of the UN Security Council Monitoring Group on the increase of foreign groups in Afghanistan is not true," Twitter post, February 6, 2022. As of June 13, 2022:
https://twitter.com/HabibiSamangani/status/1490564557349208067

Samangani, Inamullah [@HabibiSamangani] (Deputy Spokesperson for the IEA), "Maulvi Abdul Salam Hanafi expressed his gratitude for the humanitarian aid provided by the United Nations and called for more attention to be paid to the people of remote districts and areas according to their needs. 5/7," Twitter post, February 8, 2022. As of June 13, 2022: https://twitter.com/HabibiSamangani/status/1491060265311686659

Samangani, Inamullah [@HabibiSamangani] (Deputy Spokesperson for the IEA), "And someone has to tell you how you can say what this girl wants with your Ban Hijab slogan? Whatever secularism may be, but from the point of view of you fake secularists, only means enmity with Islam. As its examples can be seen in different parts of the world," Twitter post, February 9, 2022. As of June 15, 2022: https://twitter.com/HabibiSamangani/status/1491474092654047234

Samangani, Inamullah [@HabibiSamangani] (Deputy Spokesperson for the IEA), "The cabinet meeting of the Islamic Emirate was chaired by the Amir al-Mu'minin Hafezullah. On the 16th of Sha'ban 1443 AH, the cabinet meeting of the Islamic Emirate was held in Kandahar province, for three days, chaired the Amir al-Mu'minin," Twitter post, March 23, 2022. As of June 15, 2022: https://twitter.com/HabibiSamangani/status/1491474092654047234

Samangani, Inamullah [@HabibiSamangani] (Deputy Spokesperson for the IEA), "China praises the positive actions of the Islamic Emirate and it is important to establish an inclusive government and a positive change in the lives of Afghan women and children. The Chinese Foreign Minister said that his country has provided urgent assistance to the people of Afghanistan and we are focusing on areas where urgent assistance is needed," Twitter post, March 24, 2022. As of June 15, 2022: https://twitter.com/HabibiSamangani/status/1506945498195841034

Samangani, Inamullah [@HabibiSamangani] (Deputy Spokesperson for the IEA), "Maulvi Abdul Salam Hanafi called interaction, coordination and cooperation important for development and said: 'The cooperation of the international community in preventing the cultivation, use and trafficking of narcotics and providing alternatives to farmers is important.' He emphasized the need for transparency in the distribution of humanitarian aid 5/7," Twitter post, March 29, 2022. As of June 13, 2022: https://twitter.com/HabibiSamangani/status/1508742664056803335

Samangani, Inamullah [@HabibiSamangani] (Deputy Spokesperson for the IEA), "As per the decree of the supreme leader of Islamic Emirate of Afghanistan (IEA), All Afghans are informed that from now on, cultivation of poppy has been strictly prohibited across the country. If anyone violates the decree, the crop will be destroyed immediately," Twitter post, April 2, 2022. As of June 15, 2022: https://twitter.com/HabibiSamangani/status/1510495607517065221?s=20&t=yzoKi32zdQo3i02H1YcJig

Samangani, Inamullah [@HabibiSamangani] (Deputy Spokesperson for the IEA), "The aggressors, investing more than ten billion dollars, failed to eliminate opium, but the Islamic Emirate will do this task just by a decree from Amir ul-Mumineen. Inshallah," Twitter post, April 3, 2022. As of June 15, 2022: https://twitter.com/HabibiSamangani/status/1510537768950243333?s=20&t=WEJFJgsBzIjCYfHwAHhOWQ

Sarwary, Bilal [@bsarwary] (freelance journalist), "Sharp increase in Opium prices across Helmand province," Twitter post, May 1, 2022. As of June 5, 2022: https://twitter.com/bsarwary/status/1520901429585760257?s=20&t=tPTK2Y9BocVTCT6cFN1_gA

Seldin, Jeff, "How Afghanistan's Militant Groups Are Evolving Under Taliban Rule," *VOA News*, March 20, 2022. As of June 13, 2022:
https://www.voanews.com/a/how-afghanistan-s-militant-groups-are-evolving-under-taliban-rule/6492194.html

Semple, Michael, "The Capture of Mali Khan," *Foreign Policy*, October 10, 2011.

Shaheen, Suhail [@suhailshaheen1] (Taliban Permanent Representative Nominee to UN and Head of Political Office, former Negotiations Team's Member), "1/2 Reserve of Da Afghanistan Bank does not belong to governments or factions but it is property of the people of Afghanistan. It is only used for implementation of monetary policy, facilitation of trade and boosting of financial system of the country. It is never intended to be," Twitter post, February 12, 2022. As of June 13, 2022:
https://twitter.com/suhailshaheen1/status/1492709569147064320

Shaheen, Suhail [@suhailshaheen1] (Taliban Permanent Representative Nominee to UN and Head of Political Office, former Negotiations Team's Member), "2/2 used for any other purpose rather than that. It's freezing or disbursement unilaterally for any other purpose is injustice and not acceptable to the people of Afghanistan," Twitter post, February 12, 2022. As of June 13, 2022:
https://twitter.com/suhailshaheen1/status/1492709570749382657

Shaheen, Suhail [@suhailshaheen1] (Taliban Permanent Representative Nominee to UN and Head of Political Office, former Negotiations Team's Member), "1/2 I met His Excellency Markus Putzel, German ambassador to Afghanistan in Doha today and discussed with him a number of issues including removal of current sanctions, needs for reconstruction and development projects in the country," Twitter post, March 22, 2022. As of June 13, 2022:
https://twitter.com/suhailshaheen1/status/1506323269028360204

Siddique, Abubakar, and Abdul Hai Kakar, "Al-Qaeda Could Flourish with New Strategy Under Taliban Rule," Radio Free Europe/Radio Liberty, September 30, 2021.

Supreme Court of Afghanistan [@SupremeCourtAfg] (official Twitter account), "The IEA leadership has approved the establishment of the counter-narcotics court," Twitter post, January 24, 2022.

"Taliban's Mullah Omar Celebrates Prisoner-Swap 'Victory,'" BBC News, June 1, 2014.

TOLOnews, "TOLOnews 6pm News—14 March 2022," video, March 14, 2022a. As of June 12, 2022:
https://www.youtube.com/watch?v=OZu3ls6XcqI

TOLOnews, "TOLOnews 6pm News—24 March 2022," video, March 24, 2022b. As of June 12, 2022:
https://www.youtube.com/watch?v=EE5KSabNSGk

TOLOnews, "TOLOnews 6pm News—01 April 2022," video, April 1, 2022c. As of June 12, 2022:
https://www.youtube.com/watch?v=bS7O-NQEtqo

TOLOnews, "TOLOnews 6pm News—06 April 2022," video, April 6, 2022d. As of June 12, 2022:
https://www.youtube.com/watch?v=Y2w1fRrFlg4

United Nations Office of Drug and Crime, *Afghanistan Opium Survey 2021—Cultivation and Production*, March 22, 2022.

U.S. Department of the Treasury, "Treasury Issues Additional General Licenses and Guidance in Support of Humanitarian Assistance and Other Support to Afghanistan," December 22, 2021.

VOA [Voice of America] Dari, "Taliban Control Huge Amount of Opium Confiscated by the Previous Regime," video, April 22, 2022. As of January 30, 2023: https://www.facebook.com/voadari/videos/1192459594900219/

Voice of Jihad (Al Emarah English), homepage, undated. As of January 3, 2022: http://www.alemarahenglish.af

Wardak, Mohammad Naeem [@IeaOffice] (Spokesman of the Political Office), "1 / 8- Amir al-Mu'menin issued a special decree on women's rights in the name of God: The Supreme Leader of the Islamic Emirate directs all officials of the Islamic Emirate, religious scholars and tribal elders to take serious measures to ensure the following rights of women: 1- The consent of adult girls is necessary during marriage," Twitter post, December 3, 2021. As of June 6, 2022: https://twitter.com/IeaOffice/status/1466668847293812737

Wardak, Mohammad Naeem [@IeaOffice] (Spokesman of the Political Office), "A number of women marched in Kabul today with the slogan (Hijab of our honor, Hijab of our adornment, Hijab of our pride). At the end of the march, they declared their full satisfaction and support for the Islamic Emirate," Twitter post, January 20, 2022. As of June 6, 2022: https://twitter.com/IeaOffice/status/1484213182403166211

Wardak Mohammad Naeem [@IeaOffice] (Spokesman of the Political Office), "Five female police officers were employed at the Passport Directorate of Maidan Wardak. The police chief of Maidan Wardak told to Bakhtar News Agency that the female police would work to facilitate the biometric process of female's passports," Twitter post, March 7, 2022.

Wardak, Mohammad Naeem [@IeaOffice] (Spokesman of the Political Office), "Islam has granted a woman the best rights of life and they must be observed and given to all the women," Twitter post, March 8, 2022.

Wasiq, Ahmadullah [@WasiqAhmadullah] (RTA Director), "His Excellency, Mr. Muttaqi expressed the readiness of the Islamic Emirate in these areas and also demanded that special attention should be paid to the solution of the refugee problem in Iran and steps should be taken to facilitate the travelers to and from the border," Twitter post, October 23, 2021. As of June 13, 2022: https://twitter.com/WasiqAhmadullah/status/1451921528640188417

Wasiq, Ahmadullah [@WasiqAhmadullah] (RTA Director), "You can listen the live speech of Mullah Hassan Akhund," Twitter audio post, November 27, 2021. (Original tweet no longer available; audio post links to https://www.facebook.com/ariananews/videos/867197190659078/)

Wasiq, Ahmadullah [@WasiqAhmadullah] (RTA Director), "Mawlawi Abdul Kabir, political deputy of the IEA, said today in a meeting with Uzbekistan's special envoy for Afghanistan: 'For recognition, we have met all the conditions of the world and the world must continue to provide humanitarian assistance to Afghans without conditions, all problems should be solved through discussions instead of pressure,'" Twitter post, November 30, 2021. As of June 13, 2022: https://twitter.com/WasiqAhmadullah/status/1465601175936450560

World Bank, "World Development Indicators," 2020.

World Bank, "Urgent Action Required to Stabilize Afghanistan's Economy," April 13, 2022.

World Food Programme, "Afghanistan Emergency," webpage, undated. As of January 11, 2023: wfp.org/emergencies/afghanistan-emergency

Yusufzai, Arshad, "Sirajuddin Haqqani, Feared and Secretive Taliban Figure, Reveals Face in Rare Public Appearance," *Arab News PK*, March 7, 2022.